Deep Learning Systems

Algorithms, Compilers, and
Processors for Large-Scale Production

Synthesis Lectures on Computer Architecture

Editor
Natalie Enright Jerger, *University of Toronto*

Editor Emerita
Margaret Martonosi, *Princeton University*

Founding Editor Emeritus
Mark D. Hill, *University of Wisconsin, Madison*

Synthesis Lectures on Computer Architecture publishes 50- to 100-page books on topics pertaining to the science and art of designing, analyzing, selecting, and interconnecting hardware components to create computers that meet functional, performance, and cost goals. The scope will largely follow the purview of premier computer architecture conferences, such as ISCA, HPCA, MICRO, and ASPLOS.

A Primer on Memory Consistency and Cache Coherence, Second Edition
Vijay Nagarajan, Daniel J. Sorin, Mark D. Hill, and David Wood
2020

Innovations in the Memory System
Rajeev Balasubramonian
2019

Cache Replacement Policies
Akanksha Jain and Calvin Lin
2019

The Datacenter as a Computer: Designing Warehouse-Scale Machines, Third Edition
Luiz André Barroso, Urs Hölzle, and Parthasarathy Ranganathan
2018

Principles of Secure Processor Architecture Design
Jakub Szefer
2018

General-Purpose Graphics Processor Architectures
Tor M. Aamodt, Wilson Wai Lun Fung, and Timothy G. Rogers
2018

Compiling Algorithms for Heterogenous Systems
Steven Bell, Jing Pu, James Hegarty, and Mark Horowitz
2018

Architectural and Operating System Support for Virtual Memory
Abhishek Bhattacharjee and Daniel Lustig
2017

Deep Learning for Computer Architects
Brandon Reagen, Robert Adolf, Paul Whatmough, Gu-Yeon Wei, and David Brooks
2017

On-Chip Networks, Second Edition
Natalie Enright Jerger, Tushar Krishna, and Li-Shiuan Peh
2017

Space-Time Computing with Temporal Neural Networks
James E. Smith
2017

Deep Learning Systems: Algorithms, Compilers, and Processors for Large-Scale Production
Andres Rodriguez

ISBN: 978-3-031-00641-8 paperback
ISBN: 978-3-031-01769-8 ebook
ISBN: 978-3-031-00066-9 hardcover

DOI 10.1007/978-3-031-01769-8

A Publication in the Springer series
SYNTHESIS LECTURES ON ADVANCES IN AUTOMOTIVE TECHNOLOGY

Lecture #53
Editor: Natalie Enright Jerger, *University of Toronto*
Editor Emerita: Margaret Martonosi, *Princeton University*
Founding Editor Emeritus: Mark D. Hill, *University of Wisconsin, Madison*
Series ISSN
Print 1935-3235   Electronic 1935-3243

Deep Learning Systems

Algorithms, Compilers, and

Processors for Large-Scale Production

Andres Rodriguez
Intel

SYNTHESIS LECTURES ON COMPUTER ARCHITECTURE #53

ABSTRACT

This book describes deep learning systems: the algorithms, compilers, and processor components to efficiently train and deploy deep learning models for commercial applications.

The exponential growth in computational power is slowing at a time when the amount of compute consumed by state-of-the-art deep learning (DL) workloads is rapidly growing. Model size, serving latency, and power constraints are a significant challenge in the deployment of DL models for many applications. Therefore, it is imperative to codesign algorithms, compilers, and hardware to accelerate advances in this field with holistic system-level and algorithm solutions that improve performance, power, and efficiency.

Advancing DL systems generally involves three types of engineers: (1) data scientists that utilize and develop DL algorithms in partnership with domain experts, such as medical, economic, or climate scientists; (2) hardware designers that develop specialized hardware to accelerate the components in the DL models; and (3) performance and compiler engineers that optimize software to run more efficiently on a given hardware. Hardware engineers should be aware of the characteristics and components of production and academic models likely to be adopted by industry to guide design decisions impacting future hardware. Data scientists should be aware of deployment platform constraints when designing models. Performance engineers should support optimizations across diverse models, libraries, and hardware targets.

The purpose of this book is to provide a solid understanding of (1) the design, training, and applications of DL algorithms in industry; (2) the compiler techniques to map deep learning code to hardware targets; and (3) the critical hardware features that accelerate DL systems. This book aims to facilitate co-innovation for the advancement of DL systems. It is written for engineers working in one or more of these areas who seek to understand the entire system stack in order to better collaborate with engineers working in other parts of the system stack.

The book details advancements and adoption of DL models in industry, explains the training and deployment process, describes the essential hardware architectural features needed for today's and future models, and details advances in DL compilers to efficiently execute algorithms across various hardware targets.

Unique in this book is the holistic exposition of the entire DL system stack, the emphasis on commercial applications, and the practical techniques to design models and accelerate their performance. The author is fortunate to work with hardware, software, data scientist, and research teams across many high-technology companies with hyperscale data centers. These companies employ many of the examples and methods provided throughout the book.

KEYWORDS

deep learning, machine learning, artificial intelligence, distributed training systems, inference, accelerators, processors, architectures, compilers, optimizations

To Isabel, Tomas, and David

Contents

Preface

Many concepts throughout the book are interdependent and often introduced iteratively with a reference to the section covering the concept in detail. If you are new to this field, read the introductory chapter in its entirety and each chapter's introductory section and concluding paragraph to capture some of the key takeaways. Then, go back and read each chapter in its entirety. A background in linear algebra, calculus, programming, compilers, and computer architecture may be helpful for some parts but not required. The book is organized as follow:

Chapter 1 starts with an introduction to essential concepts detailed throughout the book. We review the history and applications of deep learning (DL). We discuss various types of topologies employed in industry and academia across multiple domains. We also provide an example of training a simple DL model and introduce some of the architectural design considerations.

Chapter 2 covers the building blocks of models used in production. We describe which of these building blocks are compute bound and which are memory bandwidth bound.

Chapter 3 covers the applications benefiting the most from DL, the prevalent models employed in industry, as well as academic trends likely to be adopted commercially over the next few years. We review recommender system, computer vision, natural language processing (NLP), and reinforcement learning (RL) models.

Chapter 4 covers the training process domain experts should follow to adopt DL algorithms successfully. We review topology design considerations employed by data scientists, such as weight initialization, objective functions, optimization algorithms, training with a limited dataset, dealing with data imbalances, and training with limited memory. We also describe the mathematical details behind the backpropagation algorithm to train models.

Chapter 5 covers distributed algorithms adopted in data centers and edge devices (known as federated learning). We discuss the progress and challenges with data and model parallelism. We also review communication primitives and AllReduce algorithms.

Chapter 6 covers the lower numerical formats used in production and academia. These formats can provide computational performance advantages over the standard 32-bit single-precision floating-point, sometimes at the expense of lower statistical performance (accuracy). We also discuss pruning and compression techniques that further reduce the memory footprint.

Chapter 7 covers hardware architectural designs. We review the basics of computer architecture, reasons for the slower growth in computational power, and ways to partially mitigate this slowdown. We explain the roofline model and the important hardware characteristics for serving and multinode training. We also discuss CPUs, GPUs, CRGAs, FPGAs, DSPs, and

ASICs, their advantages and disadvantages, and the prominent DL processors and platforms available in the market or in development.

Chapter 8 covers high-level languages and compilers. We review language types and explain the basics of the compilation process. We discuss front-end compilers that transform a program to an LLVM internal representation (IR) and the LLVM back-end compiler. We also describe the standard compiler optimizations passes for DL workloads.

Chapter 9 covers the frameworks and DL compilers. We review in detail the TensorFlow and PyTorch frameworks and discuss various DL compilers in development.

Chapter 10 concludes with a look at future opportunities and challenges. We discuss the opportunities to use machine learning algorithms to advance various parts of the DL system stack. We discuss some challenges, such as security, interpretability, and the social impact of these systems. We also offer some concluding remarks.

Acknowledgments

The task of writing this book would not have been possible without the support of many individuals. At the risk of unintentionally leaving some out, I want to thank the following for their guidance, time, and help discussing and improving the content.

Thanks to Michael Morgan at Morgan & Claypool for inviting me to write this book as well as Natalie Enright Jerger, Christine Kiilerich, and C.L. Tondo for their continuous encouragement and excellent support throughout the writing and publishing process. Thanks to Parthasarathy Ranganathan for his guidance and for introducing me to Michael and Natalie.

Thanks to Cliff Young, Natalie Enright Jerger, Nikhil Murthy, Nicholas Lee, and Joanne Yuan for providing detailed comments that were invaluable in improving the entire manuscript.

Thanks to Robert Zak, Wolff Daniel Dobson, Vinay Phegade, Roman Dubtsov, Kreig DuBose, Matthew Brookhart, Chinnikrishna Kothapalli, Steven (Cap) Rogers, Dheevatsa Mudigere, Carlos Escapa, Tatiana Shpeisman, Vijayakumar (Kumar) Bhagavatula, Mourad Gouicem, AG Ramesh, Jian Hui Li, Romir Desai, Vadim Pirogov, Michael Goldsmith, Chris Browning, Gal Novik, Sherine Abdelhak, Abigail Wen, Md Faijul Amin, Tristan Webb, Marcel Nassar, Koichi Yamada, Anna Bethke, Ivan Kuzmin, Anahita Bhiwandiwalla, Mariano Phelipp, Brian Retford, Tiffany Shih, Jayaram Bobba, Edward Groden, Anthony Reina, Bin Wei, Jacek Czaja, Etay Meiri, Luke Ge, Ran Cohen, Derek Osborne, Jayarama Shenoy, Michael Greenfield, Madhusudhan Rangarajan, Eric Lin, Lei Xia, Albert Hu, Brinda Ganesh, Diego Caballero, Tatyana Primak, Naveen Mellempudi, Mohammad Ashraf Bhuiyan, Rajesh Poornachandran, Rinat Rappoport, Xin Wang, Yoann Foucher, Pujiang He, Jun Jin, Eric Gardner, Adam Straw, Scott Cyphers, Brian Golembiewski, Clayne Robison, Sangeeta Bhattacharya, Ravi Panchumarthy, Patric Zhao, Derssie Mebratu, Anisha Gartia, Shamima Najnin, Rajendrakumar Chinnaiyan, Zhenlin Luo, Chris Banyai, Vinod Devarampati, Alex Heinecke, Evarist Fomenko, Milind Pandit, Lei Shao, Yong Wu, Sameh Gobriel, Andrey Nikolaev, Nawab Ali, Bernhard Friebe, Nikhil Deshpande, Shashikant Kale, Amir Gholaminejad, Indu Kalyanaraman, and Greg Leeming for their excellent comments on portions of the book relevant to their respective expertise, the discussions to ensure correctness, or proofreading early drafts to improve clarity. Thanks to Roberto Gauna and Travis Belnap for their assistance making and improving many of the figures.

My biggest thanks goes to Mary-Kathryn for her unwavering friendship and fervent support, which made the writing of this book possible. And my second biggest thanks to our children Isabel, Tomas, and David, who inspire me, and who were patient and forgiving when Papi was busy writing and could not play. This book is dedicated to you. May the progress in artificial intelligence and deep learning systems make the world for you and your generation a better place.

Disclaimer: The views expressed in the book are my own and do not represent those of the Intel Corporation, previous employers, or those acknowledged. Details regarding software and hardware products come from publicly disclosed information, which may not represent the latest status of those products.

Errata: Please submit your comments and errata to `deep.learning.systems@gmail.com`. Thanks in advance for taking the time to contribute.

Andres Rodriguez
October 2020

CHAPTER 1

Introduction

A deep learning (DL) model is a function that maps input data to an output prediction. To improve the accuracy of the prediction in complex tasks, DL models are increasingly requiring more compute, memory, bandwidth, and power, particularly during training. The number of computations required to train and deploy state-of-the-art models doubles every ~3.4 months [DH18]. The required computation scales at least as a fourth-order polynomial with respect to the accuracy and, for some tasks, as a nineth-order polynomial [TGL+20]. This appetite for more compute far outstrips the compute growth trajectory in hardware and is unsustainable. In addition, the main memory bandwidth is becoming a more significant bottleneck; computational capacity is growing much faster than memory bandwidth, and many algorithms are already bandwidth bound.

The evolution of computational growth is driving innovations in DL architectures. Improvements in transistor design and manufacturing no longer result in the previously biennial 2× general-purpose computational growth. The amount of dark silicon, where transistors cannot operate at the nominal voltage, is increasing. This motivates the exploitation of transistors for domain-specific circuitry.

Data scientists, optimization (performance) engineers, and hardware architects must collaborate on designing DL systems to continue the current pace of innovation. They need to be aware of the algorithmic trends and design DL systems with a 3–5 year horizon. These designs should balance general-purpose and domain-specific computing and accommodate for unknown future models.

The characteristics of DL systems vary widely depending on the end-user and operating environment. Researchers experimenting with a broad spectrum of new topologies (also known as DL algorithms or neural networks) require higher flexibility and programmability than engineers training and deploying established topologies. Furthermore, even established topologies have vastly different computational profiles. For instance, an image classification model may have a compute-to-data ratio three orders of magnitude higher than that of a language translation model.

A mixture of specialized hardware, higher bandwidth, compression, sparsity, smaller numerical representations, multichip communication, and other innovations is required to satisfy the appetite for DL compute. Each 2× in performance gain requires new hardware, compiler, and algorithmic co-innovations.

Advances in software compilers are critical to support the Cambrian explosion in DL hardware and to effectively compile models to different hardware targets. Compilers are essential to mitigate the cost of evaluating or adopting various hardware designs. A good compiler generates code that runs efficiently and speedily executes. That is, the generated code takes advantage of the computational capacity and memory hierarchy of the hardware so the compute units have high utilization. Several efforts, detailed in Chapter 9, are ongoing toward making this possible.

The purposes of this book are (1) to provide a solid understanding of the design, training, and applications of DL algorithms, the compiler techniques, and the critical processor features to accelerate DL systems, and (2) to facilitate co-innovation and advancement of DL systems.

In this chapter, we introduce the fundamental concepts detailed throughout the book. We review the history, applications, and types of DL algorithms. We provide an example of training a simple model and introduce some of the architectural design considerations. We also introduce the mathematical notation used throughout parts of the books.

1.1   DEEP LEARNING IN ACTION

DL models are tightly integrated into various areas of modern society. Recommender models recommend ads to click, products to buy, movies to watch, social contacts to add, and news and social posts to read. Language models facilitate interactions between people who speak different languages. Speech recognition and speech generation advance human-machine interactions in automated assistants. Ranking models improve search engine results. Sequence models enhance route planning in navigation systems. Visual models detect persons, actions, and malignant cells in MRI and X-ray films.

Other DL applications are drug discovery, Alzheimer diagnosis prediction, asteroid identification, GUI testing, fraud detection, trading and other financial applications, neutrino detection, robotics, music and art generation, gaming, circuit design, code compilation, HPC system failure detection, and many more.

Despite their tremendous success across multiple prediction domains, DL algorithms have limitations. They are not yet reliable in some behavior prediction, such as identifying recidivism, job success, terrorist risk, and at-risk kids [Nar19]. Other areas with limited functionality are personalized assistants and chatbots.

Another limitation is in Artificial General Intelligence (AGI), sometimes referred to as Strong AI. AGI is where machines exhibit human intelligence traits, such as consciousness and self-awareness. The tentative time when machines reach this capability was coined by John Von Neumann as the singularity. There are mixed opinions in the AI community on the timing of singularity ranging from later in this century to never. Given the extremely speculative nature, AGI is not discussed further.

The adoption of DL is still in its infancy. There are simpler machine learning algorithms that require less data and compute, which are broadly adopted across industries to analyze data and make predictions. These include linear regression, logistic regression, XGBoost, and Light-GBM (do not worry if you are unfamiliar with these algorithms). The majority of the winning solutions to popular Kaggle challenges involve these computationally simpler algorithms.

Nevertheless, interest in DL is extraordinarily high, and its adoption is rapidly growing. High-technology companies with warehouse-scale computers (WSC) [BHR18], referred to hereafter as hyperscale companies or hyperscalers, use DL in production primarily for these workloads (in order of importance):

1. Recommendations (due to the monetization benefits) for personalized ads, social media content, and product recommendations.

2. Natural language processing (NLP) for human-machine interaction by virtual assistants (Alexa, Siri, Cortana, G-Assistant, and Duer) and chatbots/service-bots, to combat language toxicity, for language translation, and as a preprocessing step to a recommendation workload.

3. Computer vision for biometrics, autonomous driving, image colorization, medical diagnosis, and art generation.

Recommender topologies are critical to several hyperscalers; they are more closely tied to revenue generation than computer vision and NLP topologies. The overall number of servers in data centers dedicated to recommenders is likely higher than NLP and computer vision. For instance, at Facebook, recommender models account for over 50% of all training cycles and over 80% of all their inference cycles [Haz20, NKM+20].

Computer vision topologies are widely adopted across enterprise data centers and on client devices, such as mobile phones. When companies begin the adoption of DL, they often start with computer vision topologies. These topologies are the most matured and provide significant gains over non-DL approaches. Given that several open-source datasets are available in this area, the overwhelming majority of academic papers focus in computer vision: 82%; compared to 16% for NLP and 2% for recommenders due to limited public datasets [Haz20].

Model training and serving have different requirements. Training can be computationally intensive. For instance, the popular image classification ResNet-50 model requires about 1 Exa (10^{18}) operations and is considered small by today's standards [YZH+18]. Training the much larger Megatron-LM model requires 12 Zetta (12×10^{21}) operations [SPP+19]. Other models, such as some recommenders, have unique challenges often not only requiring high compute but large memory capacity and high network and memory bandwidth.

During the training process, multiple samples are processed in parallel, improving data reuse and hardware utilization. Except for memory capacity bounded workloads, most large model training happens on GPUs due to their higher (compared to CPUs) total operations per second, higher memory bandwidth, and software ecosystem.

Serving, also known as inference, prediction, deployment, testing, or scoring, is usually part of a broader application. While one inference cycle requires little compute compared to

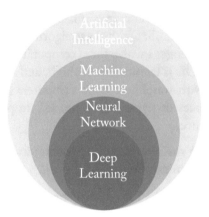

Figure 1.1: Deep learning is a subset of neural networks, which is a subset of machine learning, which is a subset of artificial intelligence.

training, the total compute spent on inference on a given model dwarfs that of training over the entire life span of the model.

Serving is typically latency bounded. Product recommendations, search results, voice assistant queries, and pedestrian detection in autonomous vehicles require results within a prespecified latency constraint. Thus, during serving, only one or a few samples are typically processed to meet the latency requirement. Effectively parallelizing the serving computations for one data sample across a large number of cores is challenging. For this reason, GPUs (and CPUs to a lesser extend) suffer from poor compute utilization during serving. There is an opportunity for hardware architects to design better low-latency processors and minimize idle compute cycles, detailed in Chapter 7.

Serving in data centers typically happens on CPUs due to their higher availability, higher core frequency, and higher compute utilization for small batches. Given the parallelization challenges when using one data sample, fewer faster cores in a CPU may be advantageous over many slower cores in a GPU. Using more cores can further reduce the latency at the expense of lower core utilization (due to the core-to-core communication overhead). However, as models grow and require more compute, some companies are transitioning to GPUs or experimenting with dedicated processors for inference. In addition, low power (smaller) GPUs or GPUs with virtualization reduces the number of cores allocated to a workload, which improves core utilization.

1.2   AI, ML, NN, AND DL

The terms artificial intelligence (AI), machine learning (ML), neural network (NN), and deep learning (DL) are often used interchangeably. While there are no agreed-upon standard definitions, the following are common and captured in Figure 1.1.

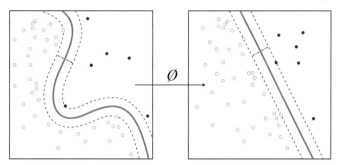

Figure 1.2: A machine learning algorithm maps the input data to a space or manifold where the data can be classified with a linear classifier. Source: [Wik11] (CC BY-SA 4.0).

AI is any program or system that can learn, act, or adapt. The recent popularity of AI comes from advances in ML algorithms, specifically in DL. An ML model is a program that learns a function that maps the input data (or features extracted from the input data) to a desired output. Geometrically, this mapping is from a vector space where the data is not linearly separable to a vector space where the data is linearly separable, as illustrated in Figure 1.2. These vector spaces are formally known as Hilbert spaces or manifolds. The mapping function or statistical performance (accuracy) of the model usually improves with more data.

NN models, also called artificial neural networks (ANNs), are typically composed of simple nonlinear functions, called layers, stacked together to represent complex functions or mappings. Stacking multiple linear functions results in one linear function that can be represented with one layer, and would negate the benefit of multilayer mappings. Thus, the need for nonlinear functions. DL models, sometimes called deep neural networks (DNNs), are NN models with more than three layers and are end-to-end differentiable. Traditional machine learning (non-NN ML) models and NN models with 1–3 layers are also called shallow models.

A difference between traditional ML and most of DL is traditional ML relies on domain experts to specify key features to use for the given task. In contrast, DL typically learns these features at the expense of requiring more data and compute. For decades, computer vision experts spent significant efforts studying image features to improve detection [FGM+10]. DL practitioners with limited computer vision expertise demonstrated that NNs were able to learn features with increasing complexity at each layer and outperform state-of-the-art techniques in image detection and classification tasks [KSH12].

DL models are particularly advantageous, although requiring much more data and compute, over traditional ML models for workloads where the relationship between features cannot be reasonably approximated, such as with visual, text, or speech data. Traditional ML models continue to be popular with tabular or structured data where the feature relationships can be approximated, for instance, using a Bayesian model to encode the hierarchical relationships manually (do not worry if you are unfamiliar with Bayesian models) [DWO+19].

1.3   BRIEF HISTORY OF NEURAL NETWORKS

NNs were popularized in the 1960s and used for binary classification. Their popularity diminished in the 1970s when NNs did not deliver on the hype. Interest in NNs increased in the mid-1980s when the backpropagation algorithm (detailed in Chapter 4) was rediscovered, facilitating the training of multilayer NNs to learn more complex classifiers. In the mid-1990s, most of the AI focus shifted toward support vector machines (SVMs), a class of ML algorithms with theoretical performance bounds. The NN community refers to the 1970s as the first AI winter and the mid-1990s to early 2000s as the second AI winter due to the limited funding of and progress in NNs (these should be called NN winters since AI is bigger than NNs).

During the past decade, there has been a revived interest as NN have vastly outperformed other techniques, particularly for vision and natural language processing tasks. This recent success is due to faster and cheaper hardware, more massive datasets, improved algorithms, and open-source software [SAD+20]. Researchers from competing companies often publish their algorithms and training methodologies (but typically not their trained models or datasets); thus, they build on each other's knowledge and accelerate progress.

1.4   TYPES OF LEARNING

ML algorithms usually fall into one of four learning types or categories: supervised, unsupervised, semi-supervised, and reinforcement learning (RL), as shown in Figure 1.3 and discussed below. Despite the names, all these learnings are "supervised" in that they required a human to explicitly define the cost function that determines what is good or bad. Note that a different way to categorize ML algorithms is as discriminative or generative. A discriminative algorithm learns to map the input data to a probability distribution. A generative algorithm learns statistical properties about the data and generates new data.

1.4.1   SUPERVISED LEARNING

Supervised learning is the most common type used in industry due to the monetization advantages and it is the primary, but not exclusive, focus of the models presented in this book. Supervised learning uses annotated or labeled data for training, meaning the ground truth or the desired output of the model for each data sample in the training dataset is known. Training involves learning a function that approximately maps the input to the desired output. The function can be a regression or a classification function. Regression functions have a numerical or continuous output, such as the price of a house (the input data would be features of the house, such as house size and local school rating). Classification functions have discrete or categorical outputs, such as {car, pedestrian, road} (the input data would be image pixels). The majority of DL models used in industry are for classification tasks. Figure 1.3a shows a classification example with the learned linear decision boundaries between three different classes. The green circles, red crosses, and blue stars represent 2D features extracted from samples in each class.

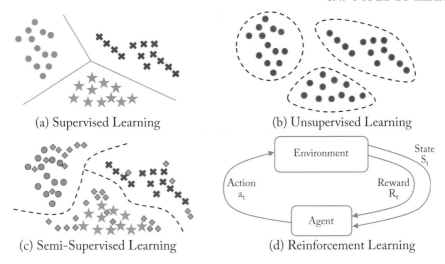

(a) Supervised Learning

(b) Unsupervised Learning

(c) Semi-Supervised Learning

(d) Reinforcement Learning

Figure 1.3: The four types of ML algorithms.

Examples of supervised learning tasks with input data and labels are (task: input data → label):

- Image classification: pixels → class label of the object in an image

- Image detection: pixels → bounding box around each object in an image and the class label of those objects

- Recommender system: shopping history, IP address, products → product purchased

- Machine translation: sentence in the source language → sentence in the target language

- Speech recognition: sound waves → written text

- Speech generation or text-to-speech (TTS): written text → sound waves

- Regression analysis: house size, local school rating → price of the house

1.4.2   UNSUPERVISED AND SELF-SUPERVISED LEARNING

Unsupervised learning learns patterns in unlabeled data. Figure 1.3b shows a clustering example with the learned clusters on unlabeled data. Self-supervised learning is a subset of unsupervised learning and includes learning embeddings and predicting missing words or pixels in text or images. For instance, each word in a 10,000-words-language can be represented as a 10,000-dimensional vector of all zeros except for a one at the index of the particular word. This vector is called a one-hot vector, shown in Figure 1.4. Self-supervised learning models can learn to map

Figure 1.4: One-hot vector. All entries are zero except a one at the vector entry corresponding to the word.

this sparse vector to a small and dense vector representation. Other examples are learning dense vector representations for persons in a social network and products in a large catalog. These dense vector representations are often the inputs into a supervised learning model.

1.4.3   SEMI-SUPERVISED LEARNING

Semi-supervised learning combines techniques from supervised and unsupervised learning. Figure 1.3c shows a small labeled dataset augmented with a much larger unlabeled dataset to improve (over the supervised learning algorithm) the decision boundaries between the classes. While most of the past decade's success has been with supervised learning, semi-supervised learning is a promising approach given the massive amounts of unlabeled data generated each day. Moreover, to draw inspiration from human learning, children appear to learn using mostly unlabeled data. However, adoption in industry is limited.

1.4.4   REINFORCEMENT LEARNING

RL is used to teach an agent to perform certain actions based on rewards received after a set of actions. The agent's goal is to maximize the rewards. Figure 1.3d depicts an agent interacting with the environment. The agent gets a reward based on the outcome of a given action. There are three types of RL algorithms: Q-learning, policy optimization and model-based, detailed in Section 3.4.

1.5   TYPES OF TOPOLOGIES

A *topology* is a computation graph that represents the structure or architecture of a NN, as shown in Figure 1.5. The nodes represent operations on tensors (multidimensional arrays), and the edges dictate the data flow and data-dependencies between nodes.

The main types of topologies used in commercial applications are multilayer perceptrons (MLPs), convolution neural networks (CNNs), recurrent neural networks (RNNs), and transformer networks. These topologies are introduced below and detailed in Chapter 3. Other types of topologies common in research with some adoption in commercial applications are graph neural networks (GNNs), adversarial networks (ANs), and autoencoders (AEs). Bayesian neural networks (BNNs) and spiking neural networks (SNNs) are limited to research.

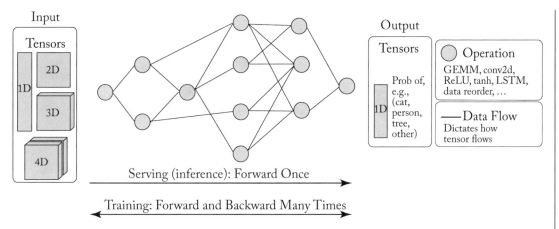

Figure 1.5: A computation graph takes a tensor input and produces a tensor output. Serving involves typically one forward propagation. Training involves numerous forward and backward iteration cycles.

1.5.1   MULTILAYER PERCEPTRON

A feedforward neural network (FFNN) is a directed acyclic graph (DAG) with an input layer, an output layer, and one or more layers in between called hidden layers. The nodes in the input layer have no parent nodes, and the nodes in the output layer have no children nodes. The inputs are either learned or extracted from the data. These models are the most widely used by hyperscalers, in particular (but not exclusively), for recommender systems.

An MLP is a vanilla FFNN with affine layers, also called fully connected layers, with each layer followed by an activation function. An affine layer is composed of units that linearly combine the weighted outputs or activations from the previous layer plus a bias (the bias is considered another weight). Using multiple layers enables the MLP model to represent complex nonlinear functions [HSW89]. Geometrically, an MLP model attempts to map one vector space to another vector space, where the data is linearly separable, via multiple nonlinear transformations, as shown in Figure 1.2. In this new manifold, the last FFNN layer is a linear classifier that separates most of the data from different classes.

Figure 1.6 shows a four-layer MLP used for digit classification with Layer 0 having $D^{(0)} = 784$ units corresponding to each pixel (the input image has 28×28 pixels), Layers 1 and 2 are hidden units having $D^{(1)} = 128$ and $D^{(2)} = 32$ units, and Layer 3 having $D^{(3)} = 10$ units corresponding to the 10 possible digits, where $D^{(l)}$ is the number of units or dimensions of Layer l. In Section 1.6, we detail how to train this model. In practice, a CNN model is a better choice for digit classification; we use an MLP model to introduce this type of topology with a simple example.

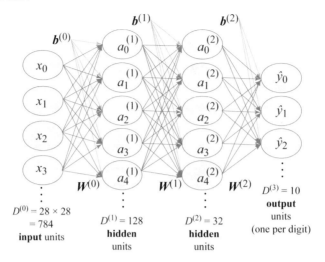

Figure 1.6: An MLP with four layers: the input layer, two hidden layers, and the output layer. This model maps the 784 pixel values to a probability distribution over 10 possible classes.

1.5.2   CONVOLUTIONAL NEURAL NETWORK

A CNN is a special type of FFNN widely used for computer vision applications, such as image classification, image detection, image similarity, semantic segmentation, human pose estimation, action recognition, and image feature extraction. Commercial applications include facial recognition, visual search, optical character recognition for document scanning, X-ray tumor detection, drug discovery, and MRI analysis. Figure 1.7 shows the input to a CNN and the output activations at each layer. Convolutional units are explained in detail in Section 2.3, and various CNN models used in production are discussed in Section 3.2.

CNNs are also used as image feature extractors; the output of one of the last layers (usually the second-to-last layer) is used as a feature vector representing the image. This vector becomes the input to other algorithms, such as an RNN to generate a textual description of the image, or to a reinforcement agent learning to play a video game, or to a recommender system that uses visual similarity.

1.5.3   RECURRENT NEURAL NETWORK

An RNN is a directed graph with nodes along a temporal or contextual sequence to capture the temporal dependencies. RNN models are used with sequential data common in language tasks and time-series forecasting. Commercial applications include stock price forecasting, text summarization, next-word recommendation, language translation, simple chatbot tasks, image description generation, speech recognition, and sentiment analysis.

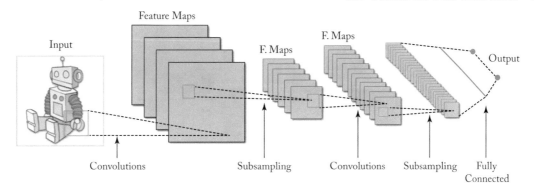

Figure 1.7: **A CNN model with several layers maps the input image to a probability distribution across multiple possible labels. Source: [Wik15] (CC BY-SA 4.0).**

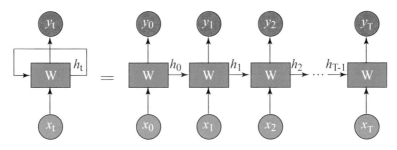

Figure 1.8: An RNN topology can be represented as an FFNN topology with the same weights **W** across all the layers.

The RNNs inputs and outputs can vary in length, unlike in MLP and CNN models. For instance, in machine translation, the input and output sentences have a different number of words. An RNN can be unrolled and represented as an FFNN sharing the same weights across the layers, as shown in Figure 1.8.

RNN models can be stacked with multiple layers and also bidirectional, as shown in Figure 1.9. The main building block of an RNN model is a recurrent unit that captures a representation or "memory" of the aggregated relevant data from previous steps. There are three main types of RNN models depending on the type of recurrent units they use: vanilla RNN, LSTM, and GRU units, detailed in Section 2.5. In the literature, the term *RNN* denotes either a *vanilla RNN* or, more broadly, these three types of models. In this book, when referring to a vanilla RNN model, we explicitly use the term vanilla RNN to prevent confusion. LSTM and GRU models are usually favored over vanilla RNN models for their superior statistical performance.

RNN models have two significant challenges: (1) capturing the dependencies in long sequences and (2) parallelizing the computation (due to the sequential nature where the output at a

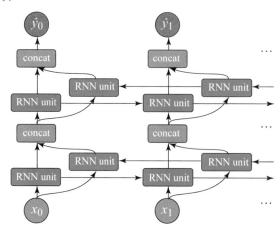

Figure 1.9: A bidirectional RNN model with two layers.

given timestep depends on the previous timestep). Using attention layers, detailed in Section 2.8, mitigates these challenges. Concatenating multiple sequential outputs from the first layer in the stack and passing those as inputs to the second layer in the stack improves the computational efficiency in a model with multiple layers [HSP+19].

1.5.4   TRANSFORMER NETWORKS

A transformer model learns how various parts of the input affects the output using an attention module. These models are also called attention-based models and have gained wide adoption for language tasks with similar applications to RNNs. They mitigate the challenges with RNNs discussed in the previous section at the expense of additional computations. The attention module can capture dependencies across long sequences. A transformer model consumes the entire sequence at once and uses multiple FFNNs in parallel together with attention modules to learn a set of weights corresponding to the influence between inputs and outputs [VSP+17]. For instance, in machine translation, the attention weights capture how each word in the output (target) language is influenced by both the neighboring words and the words in the input (source) language. The attention module is explained further in Section 2.8, and various transformer-based models used in production are discussed in Section 3.3.

1.5.5   GRAPH NEURAL NETWORK

NNs operate on data organized as vectors with MLPs, as grids or lattices with CNNs, and as sequences or chains with RNNs and Transformers. A GNN is a generalization of an MLP, CNN, RNN, and Transformer that operates on graphs rather than tensors, as shown in Figure 1.10. A graph is composed of nodes (also known as vertices) and edges representing the relation between the nodes. GNN nodes learn the properties of the neighboring nodes. Graphs are

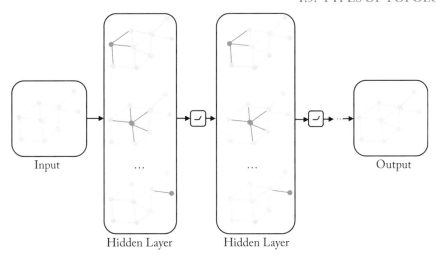

Figure 1.10: A GNN operates on graphs rather than tensors. This GNN has four layers, an input, output, and two hidden layers. Based on [Jad19].

common in many applications, such as in social networks to represent persons and their connections, in molecular biology to represent atoms and bonds, in recommender systems to represent users, items, and ratings, in telecommunications to represent networks, and in drug discovery to represent the compound structure and protein-enzyme interactions. Graphs of graphs are also common; one example is web document classification with a graph of web documents where the edges are the hyperlinks, and each node is a graph with XML-formatted elements for each document. GNNs provide the structure to learn and make predictions on graphs, often with sparsely labeled data. Given the sparse representation of the adjacency matrix in GNNs, it is beneficial to advance work in nonsequential memory access retrieval to accelerate GNNs.

GNNs were introduced in 2009 and have recently seen astronomical growth in academia [SGT+09]. Given the many real-world graph applications, rapid growth in the industry over the next few years is expected. Large-scale recommender systems, such as Pinterest's PinSage, already use GNNs [YKC+18]. Hyperscalers are developing platforms, such as Alibaba's AliGraph, Microsoft's NeuGraph, and Amazon's Deep Graph Library (DGL) to facilitate GNN industry adoption [ZZY+19, MYM+19, WVP+19]. PyTorch Geometric (PyG) is primarily targeting academic research [FL19].

1.5.6   ADVERSARIAL NETWORK

An AN or a generative adversarial network (GAN) consists of two subnetworks: a discriminator and a generator, as shown in Figure 1.11 [GPM+14]. During training, they compete in a minimax game. The generator learns to generate raw data with corresponding statistics to the training

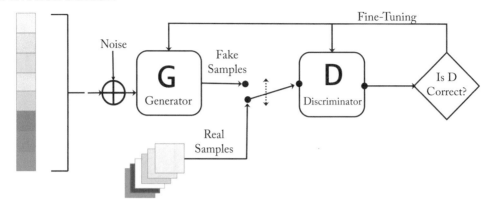

Figure 1.11: A generative adversarial network has a discriminator and a generator network that compete with each other. Based on [Gha17].

set. The discriminator evaluates the candidates as authentic or synthetic (generated). The generator's objective is to increase the error rate of the discriminator. It generates data to fool the discriminator into classifying it as authentic. The discriminator is initially trained with a training dataset. Then it is tuned as it competes with the generator. As the model trains, the generated data becomes more authentic-like, and the discriminator improves at recognizing synthetic data.

Yann LeCun, likely the most prominent DL scientist, described GANs as "the coolest idea in machine learning in the last twenty years" [Lec16]. GANs were initially proposed for unsupervised learning and now they are used across all types of learning. Wasserstein GAN (WGAN) improves the stability of learning the model, and Weng provides a detailed explanation of the mathematics used in WGAN [ACB17, Wen17].

GANs are also used for model physics-based simulations in particle physics and cosmology, reducing the simulation time by orders of magnitude [PdO+18, RKL+18]. Section 3.2.5 discusses various GANs use for image generation.

1.5.7 AUTOENCODER

An AE is a class of unsupervised learning topology that learns a low-dimensional representation (an encoding) of the input. The AE learns to reconstruct the input data in the output layer and uses the output of the bottleneck layer (usually the middle-most layer) as the low-dimensional representation. The number of units typically decreases in each layer until the bottleneck layer, as shown in Figure 1.12.

AEs are used (1) as a compact (compressed) representation of the input data; (2) as a preprocessing step to a classification problem where the data is first encoded and then passed to the classifier; (3) in a data matching problem by comparing the encoding of two data samples; (4) to

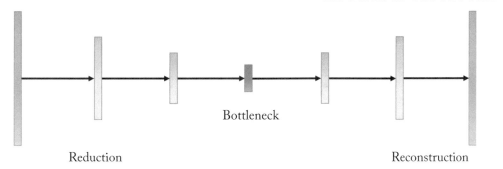

Figure 1.12: An autoencoder learns to reconstruct the input data in the output layer. The output of the bottleneck layer is often used as a low-dimensional representation of the input.

denoise data by learning a mapping from a noisy input to a clean output; and (5) as a generative model to generate data using the decoder (known as a variational autoencoder (VAE)).

1.5.8   BAYESIAN NEURAL NETWORKS

A BNN combines the strength of a NN and a Bayesian model to estimate the uncertainty of a NN prediction [Nea95]. Typical NNs use single value weights, whereas BNNs use a probability distribution over each weight; that is, BNNs provide an estimate of each weight's uncertainty, which can be used for performance guarantees and to improve interpretability. However, analytically computing and updating each distribution is prodigiously expensive. Approximating the prior and posterior distribution is an active area in research with variational inference as a common algorithm (discussed elsewhere) [BCK+15]. Despite their popularity in academia, due to their current lack of adoption in production, BNNs are not covered further.

1.5.9   SPIKING NEURAL NETWORKS

An SNN is inspired by the way natural neurons transmit information using spikes [NMZ19]. SNNs represent a whole different class of NNs differentiated by their local learning rules and are often not included in DL literature. The primary advantage of SNNs is the potential for lower power consumption using a specialized hardware known as a neuromorphic processor, such as Intel's Loihi, IBM's TrueNorth, and aiCTX's Dynamic Neuromorphic Asynchronous Processor (DYNAP) processors [RJP19]. SNNs are currently not used in production due to their inferior statistical performance and limited applications compared to other types of NNs, and therefore are not discussed further.

1.6   TRAINING AND SERVING A SIMPLE NEURAL NETWORK

A NN topology consists of the number of layers, units in each layer, and activation functions per layer. Training a model requires selecting and tuning the following hyperparameters: the NN topology, the methodology to initialize the weights, the objective function, the batch size, and the optimization algorithm and corresponding learning rate (LR). Note that in the DL literature (and in this book), *hyperparameters* are the knobs tuned by the data scientist, and *parameters* are the model weights. The type of topologies used across various workloads are discussed in Chapter 3, and the training steps are introduced below and detailed in Chapter 4. Preparing the training dataset and training with imbalanced datasets where the training samples are not evenly distributed among the classes are discussed in Section 4.5, and methods that may help identify some biases in training datasets are discussed in Sections 10.4 and 10.5. Software libraries like TensorFlow and PyTorch facilitate the training and serving of NNs and are discussed in Chapter 9. Distributed training across multiple nodes can reduce the total time-to-train (TTT), is detailed in Chapter 5.

A training system aims to reduce the time to train without sacrificing accuracy. A serving or inference system aims to maximize the throughput constrained by a latency requirement. Product recommendations, search results, voice assistant queries, and pedestrian identification in autonomous vehicles, require real-time (low latency) results. Typically, only one data sample or a micro-batch is used at a time to meet the particular application's latency requirement. Given the fewer computations per byte read from memory, the operational intensity or compute efficiency in GPUs and CPUs is lower in serving than in training.

A nomenclature note: in this book, a batch (sometimes called *mini-batch* in the literature) refers to a subset of training samples ranging from 1 to the entire dataset. A *full-batch* refers to a batch composed of the entire training dataset. A *micro-batch* refers to a batch with 1–8 samples. A *large-batch* refers to a batch size greater than 1,000 samples but less than the entire training dataset. A *node-batch* refers to the batch processed in a single node during distributed training, discussed in Chapter 5.

In the remainder of this section, we introduce some components of NNs and describe the training process using a simple example. The primary compute operations in training and serving a model are multiplications and additions, which are typically computed in groups and represented as matrices.

Once a topology is defined, training involves learning a good set of weight values. The training steps for supervised learning are typically as follows:

1. Initialize the weights or parameters of the model typically by sampling from a zero-mean Gaussian or uniform distribution.

2. Forward propagate a training sample or, more commonly, a batch of samples through the network to compute the output.

3. Evaluate the cost or penalty using a metric of difference between the expected outputs (known from the training labels) and the actual outputs.

4. Backpropagate the gradient of the cost with respect to each layer's weights and activations.

5. Update the weights of the model using the computed gradients.

6. Return to Step 2, or stop if the validation error is less than some threshold or is not decreasing.

During training, the dataset is processed in batches. The completion of a cycle through steps 2–6 for a batch is called an *iteration*, and each cycle through the entire training dataset is called an *epoch*. For instance, if the dataset has $1M$ samples and a batch has 100 samples, it takes $10K$ iterations to complete an epoch.

Training a model may require tens of epochs to learn a good set of weights. After training, the validation (also called out-of-sample) performance is measured using a validation dataset. The validation dataset contains labeled data not used during training and should be as similar as possible to the serving data the model encounters when deployed. The performance on this validation dataset is a good indicator of the performance in deployment and helps to determine if the model overfits the training dataset. Overfitting occurs when a model learns features unique to the training data and, therefore, does not generalize to data outside the training dataset. Regularization techniques to mitigate overfitting are discussed in Section 4.1.

During serving, the model processes a micro-batch. The data is propagated forward through the network to compute the output. Serving is also known as inference since the model is inferring the label of the data sample. Step 2 above is inference; that is, inference is a step in the training process but usually with a smaller batch size and some optimizations specific to serving.

The following example illustrates the training process. The task is to classify handwritten digits from the MNIST dataset using an MLP model [LBB+98]. Figure 1.13 shows a small subset of the 70,000 gray-scaled 28×28 pixel images in the MNIST dataset. Typically with MNIST, 60,000 images are used for training and 10,000 images are used for validation. In practice, a CNN model would be a better choice for image classification, but a simple MLP model is used to introduce some fundamental concepts.

Each layer in the MLP is composed of units (neurons) that linearly combine the weighted outputs or activations from the previous layer plus a bias weight, as shown in Figure 1.14 for one unit. The output from this affine transformation is passed to a nonlinear activation function $g(\cdot)$. An *activation function* refers to the nonlinear function, an *activation input* is the input to the activation function, and an *activation* (short for *activation output*) refers to the output of an activation function. Common activation functions are the rectified linear unit (ReLU) and variants of ReLU, the sigmoid and its generalization, the softmax, and the hyperbolic tangent (tanh), which are all detailed in Section 2.1.

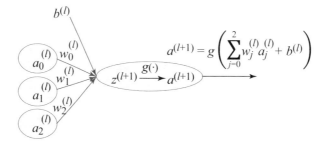

Figure 1.13: Examples from the MNIST dataset. Each digit image has 28×28 pixels. Source: [Wik17] (CC BY-SA 4.0).

Figure 1.14: A neural unit at layer $(l + 1)$ applies a nonlinear transformation or function to the weighted sum of the activations from the previous layer (l).

The MLP model used for this digit classification task, shown in Figure 1.6, has 784 units in the input layer (Layer 0) corresponding to the number of pixel values in each image. The output layer has 10 units corresponding to the probability distribution of the possible 0–9 labels. This MLP has two hidden layers with 128 and 32 units, respectively. The choice for the number of hidden layers and the number of units in each layer requires experimentation. In Section 4.5, we discuss techniques to choose an appropriate topology.

To train the model, the 28×28 image pixel values are reordered as a 784×1 vector and normalized to zero-mean and unit-norm (the benefits of normalization are explained in Section 2.6). This is the input to the NN and can be thought of as the activations of Layer 0. The input $z_i^{(1)}$ to unit i in Layer 1 is the weighted sum of the activations of Layer 0 plus a bias. The activation $a_i^{(1)}$ of unit i is a nonlinear transformation of the unit's activation input $z_i^{(1)}$:

$$a_i^{(1)} = g\left(z_i^{(1)}\right) = \max\left(0, z_i^{(1)}\right),$$

where $g(\cdot)$ is the ReLU activation function, and

$$z_i^{(1)} = \sum_{k=0}^{783} w_{ik}^{(0)} x_k + b_i^{(0)}$$

is the output of the affine transformation (also known as the activation input in Layer 1), where x_k represents the $k \in [0, 783]$th pixel value. In this example, the activation functions are ReLU for Layers 1 and 2, and softmax for the output layer. The ReLU function zeros out negative values and keeps the positive values unchanged. The softmax function is used in the output layer to map a vector of values to a probability distribution where the values are all between 0 and 1 and sum to 1. The ith output value can be computed as follows:

$$\hat{y}_i = \frac{\exp\left(z_i^{(3)}\right)}{\sum_{k=0}^{9} \exp\left(z_k^{(3)}\right)},$$

where $\hat{y}_i$ represents the probability the input image corresponds to class i. There is no bias term in a softmax layer.

This softmax output is compared with the ground truth. For this task, the ground truth is a one-hot vector with the nonzero index corresponding to the correct class label. The cross-entropy loss is:

$$-\sum_{k=0}^{9} y_k \log(\hat{y}_k),$$

where log represents the natural logarithm (log base-e), y_k is 1 if the sample belongs to class $k \in [0, 9]$ and 0 otherwise, and $\hat{y}_k$ is the model's prediction (as a probability) that the sample belongs to class k. Figure 1.15 depicts the expected and actual output for a sample image corresponding to the digit 4. In the figure, the model's output $\hat{y}$ incorrectly indicates digit 8 is the most likely inferred interpretation. Additional training iterations are needed to reduce this loss.

The gradients of the cost with respect to all the layers' activations and weights are computed using the chain rule from the last layer and moving backward layer by layer toward the first layer. Hence, the name backpropagation. The gradients provide a measurement of the contribution of each weight and activation to the cost. In practice, all of the activations for a given batch and a given layer are simultaneously computed using matrix algebra. For these computations, data scientists use software libraries optimized for the particular hardware target.

During training, the activations are saved for the backpropagation computations. Therefore, hardware for training requires a larger memory capacity than hardware for inference. The required memory is proportional to the batch size.

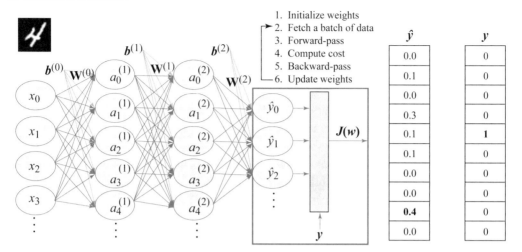

Figure 1.15: A batch of size 1 containing a sample image of the digit 4 is passed through the model. The actual output $\hat{y}$ and the expected output (ground truth) y are used to compute the cost $J(w)$. The model performs poorly in this example and predicts digit 8 with 40% probability and digit 4 with 10% probability. The cross-entropy loss is $-\log(0.1)$.

1.7   MEMORY AND COMPUTATIONAL ANALYSIS

The training process requires memory for (1) the model weights, (2) all the activations (including the batch of input data), and (3) two consecutive gradient activation layers used for gradient computations. The serving process requires memory for (1) the model and (2) two consecutive activation layers (including the input batch).

The number of weights N_w, including the biases, in the MLP model in the previous section is:

$$
\begin{aligned}
N_w &= N_{w_{L_0}} + N_{w_{L_1}} + N_{w_{L_2}} \\
&= (784 \times 128 + 128) + (128 \times 32 + 32) + (32 \times 10 + 10) \\
&= 104{,}934.
\end{aligned}
$$

This small model requires 420 KB of memory if 4 bytes are used to represent each weight. Note that in some literature, a based-2 metric is used, where a KiliByte (KiB), MiliByte (MiB), and GibiByte (GiB) represents 2^{10}, 2^{20}, and 2^{30} bytes, respectively. Thus, 420 KB is approximately 410 KiB.

The total number of activations N_a is the sum of the activations in each layer:

$$
\begin{aligned}
N_a &= N_{a_{L_0}} + N_{a_{L_1}} + N_{a_{L_2}} + N_{a_{L_3}} \\
&= (784 + 128 + 32 + 10) \times N \\
&= 954N,
\end{aligned}
$$

where N is the number of images in each batch. The size of the two largest gradient activation layers N_g required for the gradient computations, is:

$$\begin{aligned}
N_g &= N_{a_{L_1}} + N_{a_{L_2}} \\
&= (128 + 32) \times N \\
&= 160N.
\end{aligned}$$

Thus, the total memory requirement for training, using 4 bytes for each value, is:

$$\begin{aligned}
T_M &= (N_w + N_a + N_g) \times 4 \\
&= (104{,}934 + 1114N) \times 4 \\
&= 419736 + 4456N.
\end{aligned}$$

Assuming a batch of $N = 128$, the required memory for training is 1.0 MB.

The total memory requirement for inference, using 4 bytes for each value, is:

$$\begin{aligned}
T_M &= (N_w + N_a) \times 4 \\
&= (104{,}934 + (784 + 128)N) \times 4 \\
&= 419736 + 3648N.
\end{aligned}$$

Assuming a batch of $N = 1$, the require memory for inference is 424 KB.

1.8 HARDWARE DESIGN CONSIDERATIONS

The primary components in a DL platform are multitudinous multiplication and addition units, sufficient memory capacity, high memory bandwidth to feed the compute units, high inter-node bandwidth for distributed computing, and power to operate. Processing state-of-the-art models is increasingly mandating more of these components. Designing hardware requires carefully balancing these components across a huge space of numerical formats, storage and memory hierarchies, power limitations, area limitations, accuracy requirements, hardware- or software-managed caches or scratchpads, support for dense and sparse computations, domain-specific to general-purpose compute ratio, compute-to-bandwidth ratios, and inter-chip interconnects. The hardware needs the flexibility and programmability to support a spectrum of DL workloads and achieve high workload performance. In this section, we introduce some of these components and expand upon them in Chapter 7.

The core compute of training and serving are multiplications and additions. Compute is inexpensive relative to main memory bandwidth and local memory. Moore's Law continues to deliver exponential growth in the number of transistors that can be packed into a given area. Thus, the silicon area required for a set of multiply-accumulate (MAC) units is decreasing. While hardware companies often highlight the theoretical maximum number of operations (ops) per second (ops/s or OPS), the most significant bottlenecks are typically the main memory bandwidth and the local memory capacity. Without sufficient bandwidth, the overall compute

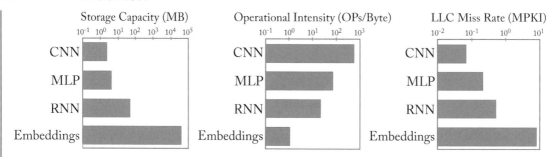

Figure 1.16: Typical CNNs, MLPs, RNNs, and embeddings differ by orders of magnitude in storage, operational intensity, and memory access irregularities. Based on [Haz20].

efficiency or utilization (the percentage of used compute cycles vs. the total compute capacity) is low for workloads bottlenecked by bandwidth (also known as bandwidth bound), and adding more compute capacity does not improve their performance. Keeping the data close to the compute can alleviate this bottleneck. In order of decreasing access time and increasing die area, the storage types are nonvolatile memory (flash memory, magnetic disk), DRAM (HBM2/E, GDDR6, DDR4, LPDDR4/5), SRAM (scratchpad, cache), and registers. DRAM is often called main memory and SRAM local memory.

The design of a balanced platform is complicated by the spectrum of workloads with diverse compute, memory, and bandwidth requirements. For instance, the CNNs, MLPs, RNNs, and embeddings used at Facebook (and similar at other hyperscalers) differ by orders of magnitude in these requirements, as shown in Figure 1.16 [Haz20]. Operational intensity is a measure of the number of operations performed per byte read from memory. The last level cache (LLC) miss rate as measured by misses per 1000-instructions (MPKI) is a standard metric to analyze the local memory (SRAM)'s efficient use and can be a metric for the irregular memory access patterns of a workload.

The numerical format is another design consideration that can impact the computational (speed) performance and statistical (accuracy) performance. Figure 1.17 shows various numerical formats, detailed in Section 6.1. A numerical representation with fewer bytes can improve the number of operations per cycle and reduce power consumption but may result in lower statistical performance. Training uses single-precision floating-point ($fp32$) with half-precision floating-point ($fp16$) and bfloat16 ($bf16$) rapidly gaining adoption. Inference uses $fp16$ and $bf16$ with 8-bit integer ($int8$) gaining adoption for some applications. A research area is developing numerical representations that can better represent values using 8 bits, such as $fp8$, discussed in Section 6.1, and can be efficiently implemented in silicon. Other techniques to reduce the memory and bandwidth requirements are increasing the sparsity and compressing the data.

A MAC unit computes the product of two values and aggregates the result to a running sum of products. The numerical format of the output (the accumulation) may be different

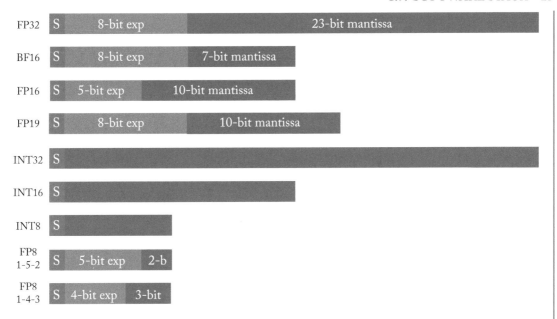

Figure 1.17: Numerical formats. Green is the sign bit. Brown are the exponent bits. Blue are the mantissa bits.

from the input. Computations involving dot products, such as in matrix multiplications and convolutions, typically use MACs. When describing MAC units, the notation used is *MAC-input-format → MAC-accumulate-format*. For instance, *int8 → int32* means the *int8* values are multiplied and accumulated as *int32* values. Accumulating values in a large numerical format mitigates numerical overflows.

Different hardware usages have different requirements. Table 1.1 shows the high-level requirements for common usages by hyperscalers: topology design, training established production models (Trn. Prod.), data center inference (Inf. DC), and edge inference (Inf. Edge). In the table, format refers to the number of bits to represent the weights and activations. Training requires more memory and bandwidth than inference to transfer and store the activations. Another use case not shown in Table 1.1 is for hardware design, which requires reconfigurable hardware (for example, FPGAs) or hardware simulators.

1.9   SOFTWARE STACK

Software is critical to a DL system. The software stack is organized as follows:

- deployment and training management systems;

- frameworks, inference engines, and graph compilers;

Table 1.1: Hardware characteristics according to usage

Usage	Format (bits)	Comp	Main Mem	Mem BW	Programmability	Internode	Interserver
Design	32	High	High	High	High	Yes	Some
Trn. Prod.	16 & 32	High	High	High	Mid	Yes	Yes
Inf. DC	16 & 8	Mid	Mid	Mid	Mid	Some	No
Inf. Edge	8	Low	Low	Low	Low	Some	No

- DNN primitive libraries and tensor compilers;

- instruction set architecture (ISA); and

- operating systems.

The primary software stack design goals are ease-of-use and high performance across various models and hardware devices.

A deployment and training management system facilitates taking a model across the pipeline stages: data preparation, topology exploration, experiment tracking, model packaging, at-scale model deployment, and retraining. The management system is designed to meet the needs of the data scientist and the infrastructure team. It provides a collaborative and secure environment, and access to the latest ML libraries, such as TensorFlow and PyTorch.

At the core of the software stack are compilers to transform the programmer's high-level code into executable code that runs efficiently on a target device. Frameworks and inference engines (IEs), such as TensorFlow, PyTorch, OpenVINO, and TensorRT, provide a high-level abstraction to the operators used across DL models. They use graph optimizers (either built-in or external) to optimize the model. The framework's scheduler relies on low-level DL and Math libraries, such as oneDNN (formerly called Intel MKL-DNN), Nvidia cuDNN, Eigen, or OpenBLAS, or in tensor compilers for optimizations to standard DL functions. Frameworks also have a code generation path to supplement these libraries with other compilers, such as LLVM.

The ISA defines the operators, data types, and memory management for an abstract computer architecture. A particular implementation of an ISA is called a *microarchitecture*. For instance, Intel and AMD CPUs use the x86 or x86-64 ISA across different microarchitecture implementations and CPU generations. Programs are binary compatible across all microarchitecture implementations of a particular ISA. Different microarchitectures can have different properties that can affect their performance, such as instructions latencies and cache hierarchies. A specific microarchitecture can be available in various flavors with different frequencies and cache sizes.

The operating system (OS) manages all the hardware and software in a compute device; it allocates hardware resources, such as compute and memory, to the software applications. An overview of operating systems is beyond the scope of this book.

Chapter 8 introduces programming languages and compiler techniques, and Chapter 9 details the prevalent DL graph and tensor compilers. Chapter 10 highlights higher-level platforms used by hyperscalers to manage training and deployment.

1.10   NOTATION

This section references the notation used throughout this book to represent input data, labels, weights, affine transformations, activations, and outputs. Recall that the compute operations in training and serving boil down to multiplications and additions. Linear algebra is used to represent groups of multiplications and additions as a single matrix-matrix or matrix-vector or vector-vector operation. While helpful, a background in linear algebra is not required; the reader can overlook the equations without a significant impact on the other parts of the book.

In DL literature, the output from an affine transformation can be equivalently represented as either

$$z_j^{(l+1)} = \sum_{i=0}^{D^{(l)}-1} w_{ji}^{(l)} a_i^{(l)} + b_j^{(l)}$$

or as

$$z_j^{(l+1)} = \sum_{i=0}^{D^{(l)}} w_{ji}^{(l)} a_i^{(l)},$$

where the bias term $b_i^{(l)}$ is included in the second equation as an additional weight with a corresponding $a_{D(l)}^{(l)} = 1$ appended to the activations. In this book, we use the first notation and explicitly represent the bias separately. The addition notation used is as follows:

- Superscripts in parenthesis means layer number

- Superscripts in brackets means sample number

- Subscript represents indices in matrices or vectors

- Bold-font lowercase represents vectors

- Bold-font uppercase represents matrices

- $\mathbf{x}^{[n]}$ and $\mathbf{y}^{[n]}$: input features and expected output (ground-truth), respectively, for the nth sample

- $(\mathbf{x}^{[0]}, \mathbf{y}^{[0]}), \ldots, (\mathbf{x}^{[N-1]}, \mathbf{y}^{[N-1]})$: training data with N samples

- $y \in \{0, 1\}$: for binary classification

- $\mathbf{y} \in \Re^M$: typically a vector with a one at the entry corresponding to its class assignment and zeros everywhere else for M-nary ($M > 2$) classification

- $\hat{\mathbf{y}} = f_{\mathbf{w}}(\mathbf{x}) \in \Re^M$: output of the model

- $D^{(l)}$: number (dimensions) of units at Layer $l \in [0, L-1]$, where L is the number of layers (note that $D^{(L-1)} = M$ for M-nary classification)

- $\mathbf{W}^{(l)} \in \Re^{D^{(l+1)} \times D^{(l)}}$: weights (not including the biases) from Layer l to Layer $l+1$, where

$$\mathbf{W}^{(l)} = \begin{bmatrix} w_{00}^{(l)} & w_{01}^{(l)} & w_{02}^{(l)} & \cdots & w_{0(D^{(l)}-1)}^{(l)} \\ w_{10}^{(l)} & w_{11}^{(l)} & w_{12}^{(l)} & \cdots & w_{1(D^{(l)}-1)}^{(l)} \\ \vdots & \vdots & \vdots & & \vdots \\ w_{(D^{(l+1)}-1)0}^{(l)} & w_{(D^{(l+1)}-1)1}^{(l)} & w_{(D^{(l+1)}-1)2}^{(l)} & \cdots & w_{(D^{(l+1)}-1)(D^{(l)}-1)}^{(l)} \end{bmatrix}$$

- $w_{ji}^{(l)} \in \mathbf{W}^{(l)}$: weight from output i in Layer l to input j in Layer $l+1$, where $i \in [0, D^{(l)}-1]$, and $j \in [D^{(l+1)}-1]$

- $\mathbf{a}^{(l)} = g(\mathbf{z}^{(l)}) \in \Re^{D^{(l)}}$: activation of Layer $l \in [0, L-1]$

- $\mathbf{a}^{(0)} = \mathbf{x}$: NN input (usually normalized)

- $\mathbf{z}^{(l)} \in \Re^{D^{(l)}}$: activation inputs to Layer $l \in [1, L-1]$

- $\mathbf{z}^{(l+1)} = \mathbf{W}^{(l)}\mathbf{a}^{(l)} + \mathbf{b}^{(l)} = [\mathbf{W}^{(l)} \ \mathbf{b}^{(l)}] \times [\mathbf{a}^{(l)}; 1]$, where $[\mathbf{W}^{(l)} \ \mathbf{b}^{(l)}]$ represents a matrix with $\mathbf{b}^{(l)}$ right appended to matrix $\mathbf{W}^{(l)}$, and $[\mathbf{a}^{(l)}; 1]$ represents a vector with a 1 bottom appended to vector $\mathbf{a}^{(l)}$

- $\mathbf{X} = [\mathbf{x}^{[0]}, \cdots, \mathbf{x}^{[N-1]}] \in \Re^{D^{(0)} \times N}$

- $\mathbf{Y} = [\mathbf{y}^{[0]}, \cdots, \mathbf{y}^{[N-1]}] \in \Re^{M \times N}$

- $\hat{\mathbf{Y}} = [\hat{\mathbf{y}}^{[0]}, \cdots, \hat{\mathbf{y}}^{[N-1]}] \in \Re^{M \times N}$

- $\mathbf{Z}^{(l)} = [\mathbf{z}^{(l)[0]}, \cdots, \mathbf{z}^{(l)[N-1]}] \in \Re^{D^{(l)} \times N}$

- $\mathbf{A}^{(l)} = [\mathbf{a}^{(l)[0]}, \cdots, \mathbf{a}^{(l)[N-1]}] \in \Re^{D^{(l)} \times N}$

- $\mathbf{Z}^{(l+1)} = \mathbf{W}^{(l)}\mathbf{A}^{(l)} + [\mathbf{b}^{(l)} \cdots \mathbf{b}^{(l)}] = [\mathbf{W}^{(l)} \ \mathbf{b}^{(l)}] \times [\mathbf{A}^{(l)}; \mathbf{1}]$

CHAPTER 2

Building Blocks

There are four main types of NN topologies used in commercial applications: multilayer perceptrons (MLPs), convolution neural networks (CNNs), recurrent neural networks (RNNs), and transformer-based topologies. These topologies are *directed graphs* with nodes and edges, where a node represents an operator, and an edge represents a data-dependency between the nodes, as shown in Figure 1.5.

A node, also called primitive (short for primitive function), layer, expression, or kernel, is the building block of a topology. While the number of functions developed by researchers continues to grow, for example, the popular TensorFlow framework supports over 1,000 operators, the number of functions used in commercial applications is comparatively small. Examples of these functions are ReLU, sigmoid, hyperbolic tangent, softmax, GEMM, convolution, and batch normalization.

There are three types of compute functions: dense linear functions (e.g., GEMM and convolution), nonlinear functions (e.g., ReLU and sigmoid), and reduction functions (e.g., pooling). A dense linear function is typically implemented as a matrix-wise operator and a nonlinear function as an element-wise operator. A reduction function reduces the input vector to one scalar value.

Matrix-wise operators are compute-intensive and (depending on the hardware and the amount of data reuse) can be compute bound (referred to as Math bound in some GPU literature). Element-wise operators are compute-light and memory bandwidth bound. The inputs to these functions are read from memory, and the results are written back to memory; there is no data reuse.

A common technique to improve the compute efficiency of a model is to fuse a compute-light element-wise operator into a compute-intensive matrix-wise operator. Thus, the intermediate results are not written to and then read from main memory. The element-wise computations happen immediately after the matrix-wise computations while the data is in the registers or the storage closes to the computing unit. Chapter 8 details this and other techniques to improve the efficiency via software optimizations.

In this and the next chapter, we follow a bottom-up approach. In this chapter, we introduce the standard primitives in popular models used at hyperscalers. In the next chapter, we discuss the actual models and applications built using these primitives. Readers that prefer a top-down approach may first read Chapter 3 to better understand the types of models and applications

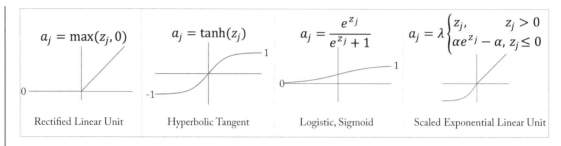

Figure 2.1: Examples of some activation functions $a_j = g(z_j)$ used in DL workloads.

before diving into the building blocks in this chapter. A review of the notation introduced in Section 1.10 can help understand the equations presented in this chapter.

2.1   ACTIVATION FUNCTIONS

An activation function is a nonlinear function applied to every element of a layer's input tensor. The most popular activation function is the rectified linear unit (ReLU). The ReLU function and its gradient are inexpensive to compute. Some models converge faster when the ReLU function is used [KSH12]. ReLU also increases sparsity which may provide computational and memory savings [GBB11].

The main drawback of ReLU is that the gradients are zero for negative activations, and the corresponding weights do not change during backpropagation. This is known as dying ReLU, and has motivated variants of ReLU, such as the Leaky ReLU (LReLU), Parametric ReLU (PReLU), Scaled Exponential Linear Unit (SELU), and the Gaussian Error Linear Unit (GELU) adopted in some attention-based models [HZR+15, KUM+17, HG16]. Another variant is ReLU6, which limits the maximum ReLU output to 6 and may improve the statistical performance when using a small numerical representation. These variants do not always result in superior statistical performance, and experimentation is required to assess the benefits.

The k-Winners-Take-All (k-WTA) activation function keeps the largest k activations in a layer and zeros out the reminder ones. A different k is typically chosen for each layer to maintain a level of constant sparsity ratio (e.g., 80%) across the layers [XZZ20].

The sigmoid and hyperbolic tangent (tanh) activation functions, shown in Figure 2.1, are commonly used in RNN models. Their main drawback is that for large positive or negative activations, the gradient is very small. During backpropagation, the product of various small gradients results in vanishing values due to the limited numerical representations in computers. This is known as the vanishing gradient problem. Variants of RNN models less prone to vanishing gradients are discussed in Section 2.5.

A benefit of the hyperbolic tangent over the sigmoid function is that it maintains a zero-mean distribution. It should be used over sigmoid except when the desired output is a probability distribution. The sigmoid function is commonly used as the activation function for the output layer in binary classification as the output is between 0 and 1 and can represent a probability.

The softmax is a generalization of the sigmoid and used for multiclass classification. It uses both element-wise and reduction operators. The output is interpreted as probability distribution with all the values between 0 and 1 and summing to 1. The ith output value can be computed as:

$$\hat{y}_i = \frac{e^{z_i}}{\sum_{k=0}^{M-1} e^{z_k}},$$

where M is the number of classes. The activation input $\mathbf{z}$ to the softmax layer is called the *logit* vector or score, which corresponds to the unnormalized model predictions, and should not be confused with the logit (sigmoid) function.

Applying the exponential function to large logits magnifies the numerical errors. Therefore, it is a common practice to subtract the maximum logit m from all the logits before using the softmax function [BHH20]. The result is mathematically equivalent:

$$\frac{e^{x-m}}{e^{x-m} + e^{y-m} + e^{m-m}} = \frac{e^x e^{-m}}{(e^x + e^y + e^m)e^{-m}} = \frac{e^x}{e^x + e^y + e^m},$$

where x, y, and m are three logits.

2.2   AFFINE

An affine transformation (also known as fully-connected, feedforward, or GEMM layer) provides a weighted sum of the inputs plus a bias. Figure 2.2 illustrates an affine transformation

$$z_j^{(l+1)} = \sum_{i=0}^{D^{(l)}-1} w_{ji}^{(l)} a_i^{(l)} + b_j^{(l)},$$

and the subsequent nonlinear activation

$$a_j^{(l+1)} = g\left(z_j^{(l+1)}\right).$$

An affine transformation can be formulated as a general matrix multiply (GEMM) for all the samples in a batch and for all the units in a layer, as shown in the last equation in Section 1.10. An affine transformation is called a linear primitive in DL literature (slightly abusing the term since a linear function should not have a bias).

Using a bias is always recommended even in large networks where a bias term may have a negligible impact on performance; removing the bias has little computational or memory savings. Note that when the affine layer is followed by a batch normalization (BN) layer (discussed in Section 2.6), the bias has no statistical impact as BN cancels out the bias.

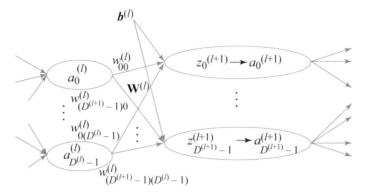

Figure 2.2: An affine layer and subsequent nonlinear function.

2.3   CONVOLUTION

Convolution kernels (commonly called filters) are widely adopted in computer vision and used with 2D images, 3D volumetric images, such as MRI scans, and 3D temporal images or video. Tasks where there is a correlation associated with the spatial or temporal proximity in the input values, such as in images, video, and spectrograms (discussed in Section 2.4), can use convolution kernels.

The term *convolution* has different meanings in the DL and signal processing literature. A convolution operator in the DL literature, and the one used in this book, is a *cross-correlation* operator between the weights and input activations (the input values to the convolutional layer). Each convolution output value is a dot product of the filter and a subset of the input. The entire convolution output is computed by shifting this filter across all the subsets in the input.

A 1D convolution using one filter follows:

$$z_i^{(l+1)} = \sum_{h=0}^{H-1} a_{h+i}^{(l)} w_h^{(l)} + b_i^{(l)},$$

where H is the length of filter $w^{(l)}$. This equation can be easily extended to a 2D convolution, which is more common in DL. Typically, multiple filters are used in each layer. Figure 2.3 illustrates K 1D convolutions and K 2D convolutions (the biases are omitted to simplify the figure).

The output is smaller if the input is not padded or if there is a stride between each filter shift. It is a common practice to extend or pad the input with zeros to enforce that the output size matches the input size (assuming the stride is 1). Another padding technique is using partial convolution, which generates a more fitting padding and is discussed elsewhere [LRS+18].

To demonstrate a 2D convolution, assume, for illustration purposes, a 6×6 gray-scaled input tensor (in practice, the input is usually much bigger) and a 5×5 filter, as shown in Fig-

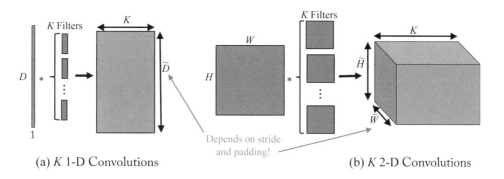

(a) *K* 1-D Convolutions   (b) *K* 2-D Convolutions

Figure 2.3: (a) K 1D convolutions and (b) K 2D convolutions. The results across all filters are concatenated across a new dimension. Thus, the output tensor of the K 2D convolutions has a depth (number of channels) of K.

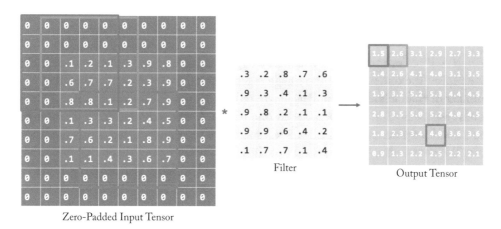

Figure 2.4: A 2D convolution operation. The top-left value in the output tensor (right) is the dot product of the values in the filter (center) with the upper left values in input tensor (left) in the red square. The input tensor is first zero-padded so the output tensor height and width dimensions equal those of the input tensor. Credit: Joanne Yuan.

ure 2.4. The input is padded with zeros to ensure the output size equals the input size. The upper left value of the 2D output array is the dot product of the 5×5 filter with the upper-left 5×5 pixels in the zero-padded input tensor (marked in red). Note that in this book and the DL literature, the dots product's definition includes the aggregated sum of the Hadamard product (element-wise multiplication) of two 2D arrays. The next output value is computed using the next 5×5 values in the input tensor (marked in green). This pattern continues across the entire input array to compute all the output values.

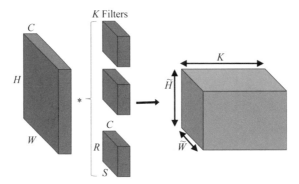

Figure 2.5: K 2D convolutions with an $H \times W$ input with C channels. Each filter also has C channels. The output tensor has K channels, each one corresponding to the convolution output of each filter with the input tensor.

An $H \times W$ color image has 3 channels (red, green, blue), also known as feature channels or tensor depth. The dimension of the image is represented as $3 \times H \times W$. The filters have the same number of channels as the input, as illustrated in Figure 2.5. Assuming K 5×5 filters with 3 channels (represented as $3 \times 5 \times 5$), each one of the K 2D convolutions is the dot product between a $3 \times 5 \times 5$ filter and all the $3 \times 5 \times 5$ subsets of the input shifted across the height and width. In 2D convolution, the filters do not shift across the depth (channels). Note that filter sizes are often described only by their height and width; the depth is inferred: it is the number of channels of the input tensor.

A convolutional layer has a bank of filters, each detecting different features in the input. To illustrate, suppose the input is a $3 \times 224 \times 224$ tensor, and the layer has a bank of 64 filters. Each filter produces one 224×224 output. Each output contains the features detected in the input by the corresponding filter. The aggregated layer output is a $64 \times 224 \times 224$ tensor, and all the filters in the next convolutional layer have 64 channels.

In practice, a convolution layer typically uses 4D input, filter, and output tensors. The usual way tensors are arranged in (1D) memory, known as the *data layout*, is as *NCHW* or *NHWC* for the input tensors, where N is the number of samples in the batch, C is the input depth (or equivalently, the number of channels or features), W is the input width, and H is the input height. The filters are arranged as *RSCK* or *KCRS*, where K is the number of filters (also known as the number of output feature channels), R is the filter height, and S is the filter width. The C in *NCWH* and *KCRS* are the same. Note that *KCRS* is sometimes denoted as *OIHW* in some literature but not in this book to avoid confusion with the H and W used for the input tensor. In the example above, the input has *NCHW* dimensions $1 \times 3 \times 224 \times 224$, the filter has *KCRS* dimensions $64 \times 3 \times 5 \times 5$, and the output has $NK\tilde{H}\tilde{W}$ dimensions $1 \times 64 \times 224 \times 224$.

The convolution is computed along seven dimensions: batch size N, output channels K, input channels C, output height $\tilde{H}$, output width $\tilde{W}$, filter height R, and filter width S. It can be implemented naively as seven `for` loops, as shown in Algorithm 2.1, where k, $\tilde{h}$, and $\tilde{w}$, represent the channel, height, and width indices of the output tensor $\mathbf{Z}$. For simplicity, the stride is assumed to be 1. There are more efficient implementations that account for a device's memory hierarchy and parallel computing capabilities [DAM+16].

Algorithm 2.1 Convolution primitive (Naive implementation)

---

init $\mathbf{z}^{(l+1)} = \mathbf{0}$
for $n \in 0, \cdots, N - 1$ **do**
  for $k \in 0, \cdots, K - 1$ **do**
    for $c \in 0, \cdots, C - 1$ **do**
      for $\tilde{h} \in 0, \cdots, \tilde{H} - 1$ **do**
        for $\tilde{w} \in 0, \cdots, \tilde{W} - 1$ **do**
          for $r \in 0, \cdots, R - 1$ **do**
            for $s \in 0, \cdots, S - 1$ **do**
$$z^{(l+1)}_{n,k,\tilde{h},\tilde{w}} \mathrel{+}= a^{(l)}_{n,c,\tilde{h}+r-1,\tilde{w}+s-1} \cdot w^{(l)}_{k,c,r,s}$$

A convolutional filter can be implemented as a GEMM by duplicating some of the input values, as shown in Figure 2.6, and converting the filter into a vector. This is called an *im2col*-based convolution implementation. This implementation enables the use of a GEMM library (which is typically well optimized). However, it comes at the expense of additional memory requirements for the input tensor and extra compute cycles to transform the input. Conversely, an affine layer can be represented as a convolution layer where C is the number of input activations, K is the number of output activations, $H = W = 1$, and $R = S = 1$ using the *NCHW* and *KCRS* data layout representations.

In addition, a convolution operation can be implemented in the Winograd domain and also in Fourier domain using the Fast Fourier transform (FFT) algorithm [HR15, LG16]. oneDNN and cuDNN support Winograd convolutions, which may be beneficial for 3×3 convolutions (the maximum theoretical gain is 2.25× over regular convolutions). However, using the Winograd transform may reduce the numeric accuracy of the computations, which can impact the overall accuracy of the model [LG16]. Sparsifying the Winograd domain (increasing the number of zero values in the Winograd domain) can lead to higher gains, but also higher accuracy loss [LPH+18]. Winograd requires additional memory and consumes more bandwidth; thus, when sufficient compute is available, the conversion overhead may result in overall slower performance. The FFT primarily benefits large filter sizes, which are uncommon in DL. Therefore, Winograd- and FFT-based convolutions are rarely used in production.

	Input	Filter	Output

Convolution:

$$\begin{bmatrix} 1 & 2 & 3 \\ 4 & 5 & 6 \\ 7 & 8 & 9 \end{bmatrix} \star \begin{bmatrix} 1 & 2 \\ 3 & 4 \end{bmatrix} = \begin{bmatrix} 37 & 47 \\ 67 & 77 \end{bmatrix}$$

Matrix Multiplication:

$$\begin{bmatrix} 1 & 2 & 4 & 5 \\ 2 & 3 & 5 & 6 \\ 4 & 5 & 7 & 8 \\ 5 & 6 & 8 & 9 \end{bmatrix} \times \begin{bmatrix} 1 \\ 2 \\ 3 \\ 4 \end{bmatrix} = \begin{bmatrix} 37 \\ 47 \\ 67 \\ 77 \end{bmatrix}$$

Figure 2.6: A convolution operation can be implemented as a matrix multiplication. In this simple illustration, the input is not zero-padded so the output dimensions are smaller than the input dimensions.

Section 3.2 introduces other variants of convolution, including the 1×1 convolution, group convolution, and depthwise separable convolutions, when discussing influential computer vision topologies.

2.4 POOLING

Pooling or subsampling reduces the size of the input tensor across the height and width, typically without affecting the number of channels. Pooling often follows a convolutional layer. The common implementation, known as *max pooling*, is to select the maximum value in a small region. A 2D pooling layer uses 2×2 nonoverlapping regions and reduces the tensor size by a factor of 4, as illustrated in Figure 2.7.

The main benefit of pooling is that filters after a pooling layer have a larger receptive field or coverage on the original input image. For example, a 3×3 filter maps to a 6×6 portion of the input image after one 2×2 pooling layer. A 3×3 filter deeper in the model, after five convolutional and pooling layers, maps to a 96×96 (note that $3 \times 2^5 = 96$) portion of the input image and can learn more complex features. Another benefit of pooling is that it reduces the number of operations.

Other forms of pooling include *average pooling*, *stochastic pooling*, and *spatial pyramid pooling* (SPP). Average pooling and stochastic pooling are similar to max pooling. Average pooling computes the average of the values in a small region. Stochastic pooling samples a value based on the distribution of the values in the small region [ZF13]. SPP is used after the last convolution layer to generate fixed-length outputs when the input images are not of a fixed size [HZR+15]. In Section 3.2.3, we provide an example of SPP used in a production model for image segmentation.

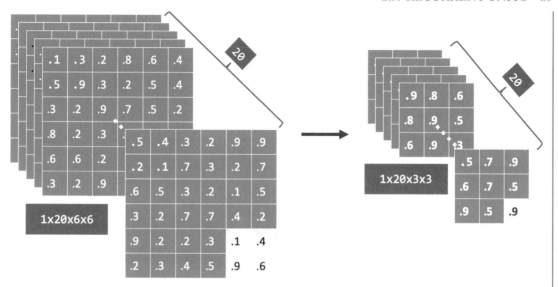

Figure 2.7: A (left) $1 \times 20 \times 6 \times 6$ tensor (in *NCHW* layout) input into a 2×2 pooling layer produces a (right) $1 \times 20 \times 3 \times 3$ tensor output. Credit: Joanne Yuan.

2.5   RECURRENT UNITS

There are three main types of recurrent units: vanilla RNN, Long Short Term Memory (LSTM), and Gated Recurrent Unit (GRU) [GSK+17, CGC+14]. Each unit has an internal vector or cell state, sometimes called the memory (not to be confused with a processor's memory). At every timestep, the input may modify this memory.

LSTM and GRU units have soft gates that control how the internal cell state is modified, as shown in Figure 2.8. These gates enable a NN to retain information for several timesteps. LSTM units have the most extensive adoption, comparable performance to GRU units, and typically statistically outperform vanilla RNN units.

LSTM and GRU units do not use activation functions between recurrent components. Therefore, the gradient does not tend to vanish during backpropagation. An LSTM and a GRU unit contain gates that allow them to control the information flow. An LSTM has a "forget" gate to flush memory cell's values, an "input" gate to add new inputs to the memory cell, and an "output" gate to get values from the memory cell. Multiplying the gate input value with the output value of a sigmoid function (the gate), which corresponds to a number between 0 and 1, implements this gating. Note the input, output, and memory cell are vectors, and each vector's value uses a unique gating value.

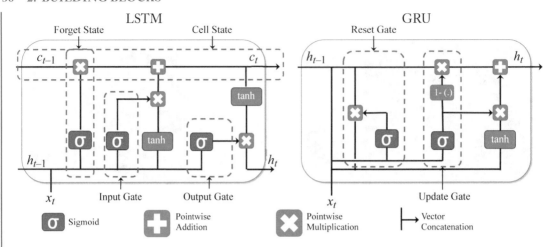

Figure 2.8: LSTM and GRU units have soft gates that control how the memory cell values are modified. Based on [Phi18].

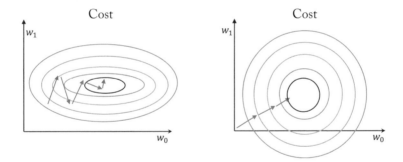

Figure 2.9: The cost space as a function of two weights for (left) unnormalized data and (right) normalized data. Each contour represents a set of weights with equal cost and the minimum is in the inner contour. Normalizing the data results in faster learning because each parameter can make a similar contribution.

2.6   NORMALIZATION

A common ML technique that improves training is normalizing the input data by subtracting the mean of the data and dividing it by the standard deviation. Normalization improves learning in single layer models as each parameter can make similar contributions to the learning, as illustrated in Figure 2.9. It is also beneficial to carefully normalize the inputs of some the layers.

The distribution of the inputs to each layer through the network can vary widely, resulting in some gradients that have little impact on the learning process. Normalizing the inputs or outputs of the activation functions improves training stability, enables the training of larger

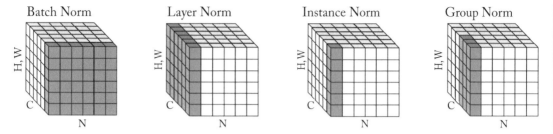

Figure 2.10: Different normalization methodologies normalize across different portions of the tensor. The tensor values colored in green are normalized by their mean and variance. Based on [WH18].

models, and results in faster convergence. The reason is that the gradient of the weights in a given layer is somewhat proportional to the magnitude of the layer inputs. Having gradients with similar magnitudes (1) reduces the effects of exploding and diminishing gradients when backpropagating through a deep network and (2) prevents some of the partial derivatives from skewing the overall gradient in a particular direction.

The most common techniques to normalize activations are batch normalization, batch renormalization, layer normalization, and group normalization, shown in Figure 2.10. In practice, we recommend using batch normalization for non-recurrent models when the batch size is greater than or equal to 32 and group normalization otherwise.

Batch normalization (BN) was a breakthrough technique enabling the training of deeper and more accurate models and is widely adopted in production [IS15]. BN can be applied to the input or output of an activation function. Based on empirical data, the latter is recommended and used in the analysis below.

The activations $\mathbf{a}^{(l)}$ in Layer l are normalized by the mean E and variance V across a batch of samples. Each BN layer has two trainable parameters: γ and β to scale and then shift the normalized activations. These parameters provide flexibility over the amount of normalization in a BN layer to maximize statistical performance. Note that data scientists can remove the bias term in the fully-connected or convolutional layer with no statistical effects as the shift term β effectively cancels out the bias term.

At the end of the training process, the mean and variance for each BN layer are computed using statistics from the entire training set or a large representative subset. These values are fixed and used during serving; they are not recomputed in each serving batch. During inference, the BN output is:

$$
\begin{aligned}
BN\left(\mathbf{a}^{(l+1)}\right) &= \gamma \frac{\mathbf{a}^{(l+1)} - E\mathbf{1}}{V} + \beta\mathbf{1} = \gamma \frac{g(\mathbf{W}^{(l)}\mathbf{a}^{(l)}) - E\mathbf{1}}{V} + \beta\mathbf{1} \\
&= g\left(\frac{\gamma}{V}\mathbf{W}^{(l)}\mathbf{a}^{(l)}\right) + \left(\beta - \frac{\gamma E}{V}\right)\mathbf{1} = g\left(\mathbf{W}'^{(l)}\mathbf{a}^{(l)}\right) + \mathbf{b}',
\end{aligned}
$$

where $g(\cdot)$ is the activation function, $\mathbf{W}' = \frac{\gamma}{V}\mathbf{W}$ and $\mathbf{b}' = (\beta - \frac{\gamma E}{V})\mathbf{1}$. That is, during inference the BN can be incorporated directly in the weights by multiplying them by $\frac{\gamma}{V}$ in the preceding convolutional or fully-connected layer, and adding the bias $\mathbf{b}'$ to the activations.

There are two drawbacks to batch normalization. First, it requires training with batches of sufficient samples (usually 32 or more) to capture adequate statistics representative of the entire training set. This requirement limits distributed training algorithms when the batch per device is small. Second, batch normalization cannot be used in recurrent layers because the statistics change with each recurrent step, but the BN parameters are shared across all steps. Batch renormalization, layer normalization, and group normalization address these drawbacks.

Batch renormalization constrains the mean and standard deviation of BN to reduce the large difference when the batch size is small [Iof17]. Batch renormalization allows training with small batches.

Layer normalization computes the mean and standard deviation across all the activation values in a layer in a data sample. Therefore, different data samples have different normalization terms [BKH16].

Group normalization is a generalization of layer normalization. It uses the mean and variance across groups of channels, rather than across all the channels [WH18]. The number of groups is a hyperparameter chosen by the user. Both of these methods also include the two trainable parameters as in batch normalization. Empirical results show group normalization works much better than BN for small batch sizes, and only slightly worse than BN for large batch sizes [WH18].

Local response normalization (LRN) square-normalizes the values using the statistics in a small neighborhood [KSH12]. LRN is not a trainable layer. It was used in older models before batch normalization gained adoption.

2.7   EMBEDDINGS

An embedding (also known as encoding or thought-vector) is a low-dimensional dense vector representation of a data sample. It is often used as the input to a language or recommender model. An embedding layer maps high-dimensional sparse data to an embedding. To illustrate, suppose a dictionary has 10,000 words. A 10,000-dimensional vector of all zeros except for a one at the corresponding index represents a word. This is called a one-hot vector. Unsupervised learning algorithms, such as word2vec or GloVe, can learn to map a corpus of words to low-dimensional dense representations [MSC+13, PSM14]. Other usages are learning dense representation for persons in a social network or products in a retail business with a large catalog. In images, the activations of the last or second-to-last layer of a CNN model are often used to embed the image.

The embeddings often demonstrate data associations, and vector embeddings of similar words are closer to each other. For instance, using their learned vector representations, $\mathbf{v}_{queen} \approx \mathbf{v}_{woman} + \mathbf{v}_{king} - \mathbf{v}_{man}$, as shown in Figure 2.11.

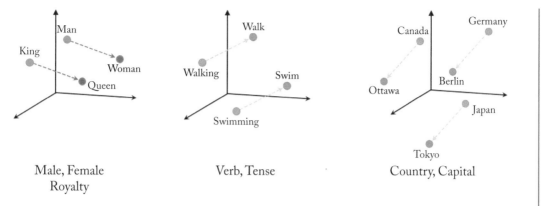

Figure 2.11: 3D word embeddings. Word embedding often capture word associations. Based on [Goo20].

2.8 ATTENTION

An attention layer learns how the input vectors influence each output vector, as shown in Figure 2.12. Some attention layers also capture how the neighboring output vectors influence each other. Attention layers are popular in language models to determine the associations between input and output tokens [VSP+17]. A token is a word, a subword, or a symbol, such as a question mark. Attention layers can be computationally expensive as each layer may require computing an attention value for each combination of input and output tokens. This additional computation may increase the serving latency in some workloads beyond what is acceptable for a given application. Nevertheless, using attention layers can improve the statistical performance.

Attention layers are also used in some recommenders to learn a context vector that captures the influence between users. They are also used in image captioning to focus the decoder on the relative parts of the image [CZZ+19, YHG+15]. Attention can improve interpretability. The attention layer may be used to observe and explain how an input feature influences a particular output.

2.9 DROPOUT

Dropout is designed to reduce overfitting and may be used in fully connected non-recurrent layers. During training, a percentage of the weights in a layer is ignored (dropped) for an iteration, as shown in Figure 2.13. At each iteration, a new set of weights is randomly ignored, which reduces overfitting by reducing cooperation between the weights. During inference, all the weights are used.

RNN-based models can use dropout after the embedding layers and in-between RNN stacks. While dropout could be used across temporal units if the same set of weights across all

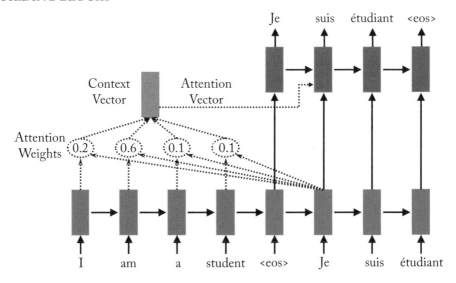

Figure 2.12: Attention-based models capture how each output token is influenced by the all input tokens. The blue and green rectangles are cells corresponding to the encoder and decoder, respectively. To generate the french word *suis*, the attention weight is the highest for the corresponding English word *am*. Based on [Syn17].

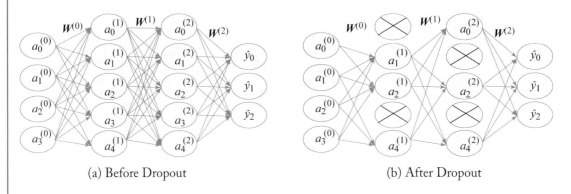

(a) Before Dropout   (b) After Dropout

Figure 2.13: (left) Original model before dropout. (right) A percentage of the weights in a layer are dropped during each training iteration.

the timesteps are dropped, it is usually not used. CNNs layers typically do not use dropout given that those layers already have few weights.

In practice, normalization techniques are preferred to reduce overfitting, and newer models do not use dropout. Based on empirical evaluations, normalization and dropout should not be jointly used as they have a negative convergence impact [LCH+19].

SLIDE is an extension to dropout [CMF+20]. In dropout, the weights are randomly dropped. In SLIDE, the weights that produce small activations are dropped. The percentage of dropped weights in SLIDE can be 90–95%, whereas in dropout it is usually 50%. Thus, only the most relevant weights are updated in each training iteration. The challenge that SLIDE addresses is predicting which weight vectors produce large activations. SLIDE uses locality sensitive hashing (LSH) to select similar vector weights to the input activation vectors.

SLIDE has a CPU affinity for two main reasons. First, LSH relies on branching, for which CPUs are well optimized. Second, the LSH tables require a large memory capacity, which also favors CPUs. There is ongoing research toward making hashing more efficient on GPUs [AFO18].

Similarly to dropout, SLIDE is primarily beneficial for fully-connected, non-recurrent layers with many units, since these are typically overparameterized and can be heavily sparsified. In particular, for the affine layer into the softmax layer in extreme classification tasks (common in recommender systems), where the softmax has hundreds of thousands of units. Similar to dropout, jointly using SLIDE and normalization is not recommended. Finally, SLIDE is relatively new and has not been adopted in production environments. Further work is needed to facilitate adoption.

In this chapter, we detailed the standard building blocks of topologies used in commercial applications and explained their purpose. These building blocks or layers have different hardware needs. Typically, embeddings layers need large memory and memory bandwidth, convolutional layers need large compute, and recurrent layers need large memory bandwidth. We introduced the concept of a graph with nodes and edges as a representation for a topology. A standard graph optimization technique, detailed in Chapter 8, is to merge dense linear nodes, such as GEMM and convolutions, with element-wise nodes, such as ReLU, to reduce memory accesses and improve performance. We recommended using batch normalization for non-recurrent layers when the batch size is greater than or equal to 32 and group normalization otherwise. Also, normalization is preferable over dropout and both should not be used jointly. In the next chapter, we discuss foundational topologies composed of these building blocks and their applications by hyperscalers.

CHAPTER 3

Models and Applications

The main types of workloads where DL models are used in production are recommender systems, computer vision, and NLP.

Recommender systems are typically the most prevalent class of models at hyperscalers, given the significant revenue they generate. Neural (DL-based) recommenders usually have two main components: an embedding layer and a NN, typically an MLP. The embedding layer maps high-dimensional sparse data to low-dimensional dense data. For instance, a movie recommender may have a dataset with 100 million users, each rating a few of the 10,000 available movies. This dataset is sparse since most users only watch a tiny subset of the movies. An embedding layer maps this sparse representation to a dense representation and then passes it to the MLP. Standard recommender models are Wide & Deep (W&D), Neural collaborative filtering (NCF), Deep & Cross Network (DCN), Deep Interest Evolution Network (DIEN), and Deep Learning Recommender Model (DLRM), covered in Section 3.1.

Computer vision models have multiple convolutional layers, often followed by a normalization function, such as batch normalization, and a nonlinear activation function, such as ReLU. Standard computer vision models are Inception, ResNet, MobileNet, UNet, Mask-RCNN, SSD, YOLO, DCGAN, and StyleGAN, covered in Section 3.2.

NLP includes natural language understanding (NLU), speech recognition, speech generation, and speech-to-speech translation. Many of these workloads use an embedding layer to map tokens or words (represented as one-hot vectors) into a low-dimensional dense representation. A one-hot vector word representation is a high-dimensional vector of all zeros except a one at the corresponding word index.

There are two main types of NLP models: RNN-based (including LSTM-based and GRU-based) and transformer-based models. RNN-based models typically have lower latency compared to transformer-based models, which require significantly more computations but typically have higher statistical performance. Standard NLP models are sequence-to-sequence, Transformer-LT, BERT, Deep Speech 2, and Tacotron, covered in Section 3.3.

Some workloads use multimodal models, where a unified output combines the last layers from multiple models. For instance, an autonomous driving vehicle may use a multimodal model that combines radar, LiDAR, and visual input. A person ID model may combine a voice ID model and a face ID model. Other workloads may use models sequentially. For instance, a recommender may use a language model as a preprocessing step in a search query, or an image descriptor may use a computer vision model followed by a language model.

We discuss RL topologies popular in academia and with some industry adoption in Section 3.4. Several applications may benefit from RL, including hardware and software design (see Section 10.1). RL topologies fall nonexclusively under three types of algorithms: Q-learning, policy optimization, and model-based. The most well-known is the model-based AlphaGo, which beat the Go world champion Lee Sedol [SHM+16].

In all these workloads, determining the best model that meets the accuracy requirements within a computational and memory budget requires experimentation. Also, each topology can require significant engineering effort to design. Recently, there are newer techniques known as *AutoML*, which includes neural architecture search (NAS), that automatically develop topologies for a particular task and hardware target, reducing the manual engineering effort at the expense of much higher computation. Section 10.1 discusses NAS techniques.

3.1   RECOMMENDER SYSTEMS TOPOLOGIES

Recommender systems, also known as recommendation engines or recommenders, are the most important type of DL algorithms at hyperscalers. They can provide significant monetization opportunities; their utility and market impact cannot be overstated. Despite their importance, only 2% of DL publications focus on recommenders, and the published results are often not reproducible [DCJ19].

Recommender systems are categorized into two approaches, and a hybrid of these two is common.

(1) Content-based systems recommend items to a user based on their profile and their user-item interaction.

(2) Collaborative filtering recommends items to a user based on the user-item interactions from similar users.

The input can be structured data, such as databases, or unstructured data, such as text and images. CNNs and RNNs can be applied to images and text, respectively, to extract features to input into a recommender. The recommended items can be ads to click, merchandise to purchase, videos to watch, songs to listen, social contacts to add, and news and social media posts to read. Recommender systems recommend items based on user features, item features, user-item ratings, and contexts, such as the time, day, and season. User-item ratings can be explicit or implicit based on user-item interaction. Implicit feedback includes the amount of time spent reading a news article, listening to a song, or viewing a clip. Details on the recent advances in context-aware recommender systems (CARS) are available elsewhere [RD19].

The total number of user-item combinations can reach quintillions, and adding context further increases that number. Netflix has around 200 million users and over 13,000 titles [Lov19]. Google Play has over 1 billion users and over 1 million apps [CKH+16]. eBay has more than 180 million buyers and over 1.3 billion listings [Don19, Pad19]. Alibaba has 2 billion products and serves as many as 500 million customers per day [Fel19]. This huge catalog results in memory size bottlenecks on the hardware platform. If every combination requires one byte, then the total user-item combinations would require 1 exabyte of memory, which is 4× more than the total storage of the largest supercomputer Summit. eBay clusters items into categories and utilizes user-category (rather than user-item) to reduce the number of combinations [Bro19].

Rather than ranking potentially billions of items, a large-scale recommender system breaks the process into two stages to meet the latency requirement and reduce the number of computations. First, a recall (candidate generator) stage selects several items that may be of interest to the user. Second, a ranking stage scores each item and selects those shown to the user [Bro19]. The recall step selects a set of items (for instance, 1000) that may be of interest to a particular user, each one represented as a vector. The dot products between the vector representing the user and each of the 1000 item-vectors are computed. The items producing the highest dot products are then recommended to the user.

Despite using a two-stage approach, a significant challenge of large-scale recommender systems is the large memory required, in particular for the embedding tables to embed users and item features. Baidu's *Phoenix Nest* online advertising recommender models can exceed 10 TB, dwarfing the capacity of a GPU. Therefore, the model is partitioned into embeddings on the CPU and the NN on the GPU on Baidu's AIBox [ZZX+19].

Content-based recommenders use features, known as metadata, for each item (such as movie genres and IMDb ratings) and recommend items based on the similarities to other items the user has liked. A profile for each user is learned based on their likes and used to evaluate the recalled items to make a recommendation. Other features can include embedding representations (using RNN, CNN, or hand-engineered features) of written reviews, movie summaries, and still images. Content-based recommenders do not use information or ratings from other users.

Collaborative filtering (CF) recommends items based on user-item interactions across multiple users. Collaborative filtering uses no metadata or domain knowledge about the items; instead, it learns all the feature vectors. This eliminates the dependency of manually chosen features at the expense of requiring more data. A *rating matrix* $\mathbf{R}$, also known as the utility matrix or user-interaction matrix, contains the ratings across various users and items. Collaborative filtering learns a user matrix $\mathbf{U}$ and an item matrix $\mathbf{V}$ composed of user and item feature vectors of equal dimension, respectively, such that the squared differences between $\mathbf{R}$ and the dense matrix $\hat{\mathbf{R}} = \mathbf{U}\mathbf{V}^T$ is minimized. This is known as matrix factorization. $\hat{\mathbf{R}}$ provides a metric of similarity between the items and users. In practice, for large rating matrices, only a subset of entries is used. The alternative least squares (ALS) algorithm can perform matrix factorization by alternating between holding constant one of the matrices and adjusting the other one to minimize the error. Singular Value Decomposition (SVD) is another commonly used matrix factorization algorithm.

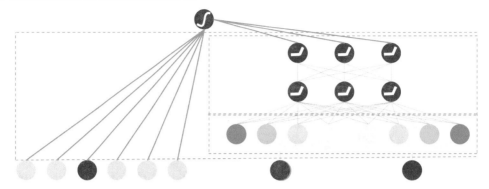

Figure 3.1: A (left) linear and (right) deep learning model. Based on [CKH+16].

Neural recommenders typically use a hybrid approach. They are trained with large datasets across multiple user and item pairs. Standard neural recommenders are Wide and Deep (W&D), Neural collaborative filtering (NCF), Deep Interest Evolution Network (DIEN), and Deep Learning Recommender Model (DLRM). GNNs are also gaining adoption for recommenders. Other recommenders include autoencoders to encode implicit ratings or feedback, GANs, and deep RL to tackle dynamic changes in items and users' preferences [LKH+18, WYZ+17, ZZZ+18].

Wide & Deep (W&D) combines the output from a linear model (referred to as wide) and a deep model, as shown in Figure 3.1 [CKH+16]. This was originally developed to improve Google Play's app recommendation. The probability the recommended app is chosen given the input vector is:

$$P(v_i = 1|\mathbf{x}) = \sigma \left(\mathbf{w}_{wide}^T \phi(\mathbf{x}) + f_{deep} \left(\mathbf{W}_{deep}, \mathbf{x} \right) + b \right),$$

where σ is the logit function, b is the bias term, $\phi(\mathbf{x})$ are the features on the linear model, $f_{deep}(\cdot)$ is the deep model, $\mathbf{w}_{wide}$ is the weight vector for the linear model, and $\mathbf{W}_{deep}$ is the set of weights for the deep model. The input vector $\mathbf{x}$ has user features (for instance, country, language, demographics), contextual features (for instance, device, hour of the day, and day of the week), and item features (for instance, app age, and historical statistics of an app). Sparse discrete high-dimensional categorical feature vectors are embedded into a low-dimensional dense representation and concatenated as one input vector into an MLP.

Similar models to W&D are the MLP model used for YouTube recommendations, which incorporates the mixture-of-experts ML technique, and the DeepFM model, which shares the input with its "wide" and "deep" parts [CAS16, ZHW+19, GTY+17]. Another similar model is Deep & Cross Network (DCN) used for ad click prediction. It applies feature crossing and, unlike W&D, does not require manually selecting the features to cross.

Neural Collaborative Filtering (NCF) is a CF-based recommender that generalizes the popular matrix factorization algorithm [HLZ+17]. A one-layer linear NN can represent matrix

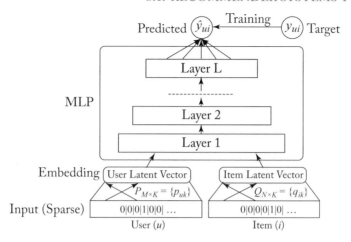

Figure 3.2: A neural collaborative filtering (NCF) model with one embedding layer for the user and one for the items, and an MLP model. Based on [HLZ+17].

factorization. NCF augments this linear NN with multiple layers, as shown in Figure 3.2, to model complex nonlinearities in the data, which improves the learned features and recommendations.

Deep Interest Evolution Network (DIEN) and **Behavior Sequence Transformer (BST)** are used in production at Alibaba's Taobao to recommend advertisements [ZMF+18, CZL+19]. They use a GRU- and a transformer-based topology, respectively, to model user behavior through time. A similar model, Recurrent Recommender Network (RRN), uses LSTM units [WAB+17].

Deep Learning Recommendation Model (DLRM) is a class of models used by Facebook. DLRM improves the handling of categorical features [NMS+19]. The dot product of pairs of embedding vectors and processed dense features are post-processed through another MLP, as shown in Figure 3.3, to predict event probabilities. Because the embedding tables are enormous, model parallelism, discussed in Chapter 6, can be used to mitigate memory constraints. Facebook also proposed using nonvolatile memory (NVM) as the primary storage medium and DRAM as a cache for commonly used embedding vectors [ENG+18].

Graph Neural Networks (GNNs), introduced in Section 1.5.5, are gaining traction for large-scaled recommender systems. Industry platforms include Pinterest's PinSage, Alibaba's AliGraph, Microsoft's NeuGraph, and Amazon's Deep Graph Library (DGL) [YKC+18, ZZY+19, MYM+19, WVP+19, FL19, Fey20].

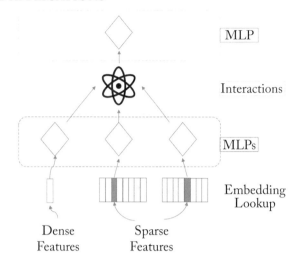

Figure 3.3: A Deep Learning Recommendation Model (DLRM) with a dense feature, multiple embedding layers for sparse features, and multiple MLP topologies. Based on [NMS+19].

3.2   COMPUTER VISION TOPOLOGIES

Computer vision topologies are the most widely adopted types of topologies across enterprise businesses. Commercial applications include image/video tagging, facial identification, autonomous navigation, video summarization, medical image analysis, and automatic target recognition with electro-optical, synthetic aperture radar (SAR), and hyperspectral imagery.

Before adopting DL, computer vision practitioners spent significant efforts in engineering features to extract from the data samples. In particular, for image classification, practitioners developed (do not worry if you are unfamiliar with these algorithms) local binary pattern (LBP), histogram of oriented gradients (HOG), and speeded up robust features (SURF) [OPM02, BTV06]. Similar to Gabor filters, these features attempt to capture the gradient information around a pixel or set of pixels, essentially acting as edge detectors [Gab46]. The features are then passed to a classifier, such as an SVM. Extracting features results in better performance than feeding the shallow classifier the raw pixels.

In 2012, Krizhevsky et al. developed the now-called AlexNet topology and decisively won the ImageNet challenge, which grabbed the attention of the computer vision and other communities [RDS+15, KSH12]. The computer vision community rapidly adopted DL techniques resulting in a lower image classification error every year (see Figure 3.4 and note the jump from 2011–2012 when AlexNet was introduced). Commercial applications no longer use AlexNet given the newer improved models, but it is mentioned given its historical significance.

CNN models learn to detect features with increased hierarchical complexity along each consecutive layer. Interestingly, the weights in the first layer of computer vision models learned

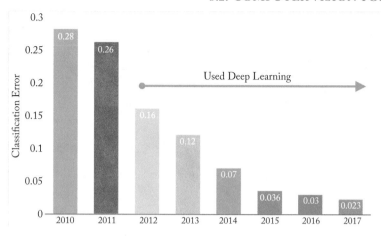

Figure 3.4: Top-5 classification error from 2010–2017 on the ImageNet-1K dataset. Since 2012 all the top results have used DL. Based on [Zis18].

similar features to those developed over decades of research, which are also similar to the features used by the mammal primary visual cortex. That is, the weights in the first layer usually become edge detectors after training the model. The second figure in Zeiler and Fergus' 2013 paper (not replicated here) shows what some of the feature maps across various layers specialized to detect [ZF13]. One difference (of many) between CNN models and the mammal visual cortex is that CNNs rely more on texture features than shape features [GRM+18]. Augmenting the training dataset by perturbing each image's texture increases the dependency on shape features and improves the model's performance.

Computer vision tasks detailed in this section are classification, object detection, semantic segmentation, verification, and image generations. Additional tasks not discussed include action recognition, image denoising, super-resolution, and style transfer.

3.2.1   IMAGE CLASSIFICATION

The task of image classification is to predict the class label of an image. A common preprocessing technique is to resize all the training and testing images to be the same size and square. Two common metrics to measure accuracy are Top-1 and Top-5. Top-1 accuracy requires that the model's top guess (corresponding to the highest logit) is the correct one. Top-5 requires that one of the model's top 5 guesses is the correct one. Top-5 helps to account for uncertainty in the labels. Figure 3.4 illustrates the Top-5 classification error from 2010 to 2017 on the ImageNet 1,000 classes ($i1k$) dataset. Since 2012 all the top results have used DL.

Key neural image classifiers include AlexNet, VGG, Inception, ResNet, DenseNet, Xception, MobileNet, ResNeXt, and NAS. These families of topologies introduce new layers dis-

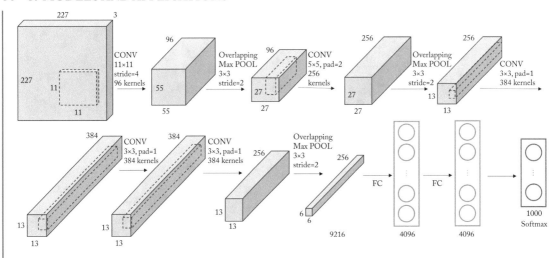

Figure 3.5: All the layers of the AlexNet topology. Based on [Has18].

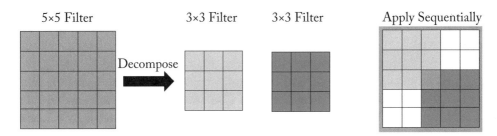

Figure 3.6: The factorization of a 5 × 5 filter into two consecutive 3 × 3 filters maintains the same receptive field.

cussed below, such as inception, residual, group convolution, and depthwise separable convolutional layers, and introduce new techniques, such as factorization.

AlexNet, shown in Figure 3.5, is similar to the 1998 **LeNet-5** topology used for digit recognition but with more layers and units. Also, AlexNet uses ReLU rather than the logit activation functions, and max pooling rather than average pooling [LBB+98].

VGG is a family of topologies similar to AlexNet but with more layers and only uses 3 × 3 convolution filters [SZ14]. VGG factorizes a 5 × 5 into two consecutive 3 × 3 layers to reduce the number of parameters, as shown in Figure 3.6. Factorization maintains the same receptive field coverage. The reduced number of parameters mitigates overfitting, which facilitates using topologies with more layers.

Inception-v1, also known as **GoogleNet**, introduced the inception module, which is composed of multiple filters of different sizes that process the same input, as shown in Fig-

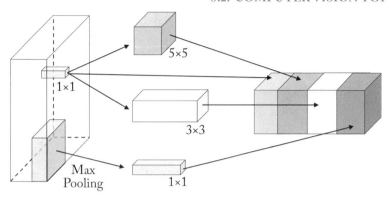

Figure 3.7: The Inception-v1 module. Different filter sizes are applied to the input tensor and the outputs are concatenated.

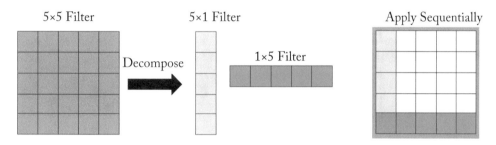

Figure 3.8: The factorization of a 5×5 filter into a 5×1 and 1×5 filters maintains the same receptive field.

ure 3.7 [SVI+15, SLJ+14]. These filters extract multilevel features, and their outputs are concatenated. The inception module popularized the usage of 1×1 filters, which modifies the number of channels. Inception replaces the fully-connected layers at the end of the topology with a global average pooling across the 2D feature map, which notably reduces the total number of parameters.

Inception-v3 introduces the factorization of an $n \times n$ convolutional filter into a $1 \times n$ followed by an $n \times 1$ filter, as shown in Figure 3.8. This factorization maintains the same receptive field and reduces the number of weights from n^2 to $2n$. Inception-v3 also adds a regularization known as *label smoothing* to the one-hot label vectors by replacing the zeros with a small epsilon value. Inception-v3, like Inception-v2 (also known as Batch-Normalization Inception), uses batch normalization [IS15].

Another technique introduced in VGG and improved in Inception-v3 is doubling the number of channels and halving the feature maps' length and width in consecutive layers. This pattern is made in one of three ways. First, convolution followed by pooling at the expense of a

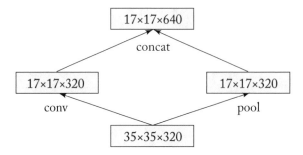

Figure 3.9: Efficient grid size reduction. The number of channels doubles and the height and width are halved.

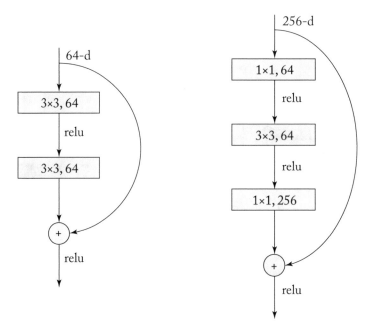

Figure 3.10: Residual layers have skip connections that bypass certain layers. (left) A residual layer with two convolutional layers. (right) A residual module reduces the tensor to 64 channels (from 256 channels) to reduce the number of 3×3 convolutions and then expands the output back to 256 channels.

convolution with a larger tensor input. Second, pooling followed by convolution at the expense of a less-expressive layer. Third (recommended), two parallel blocks: (1) a convolution block with a stride of 2 that maintains the same number of channels; and (2) a pooling layer, as shown in Figure 3.9.

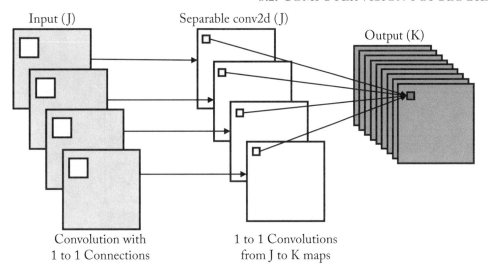

Figure 3.11: Depthwise separable convolution is a depthwise convolution, where every input channel is convolved with a different filter, followed by a pointwise convolution.

ResNet is a family of models that popularized layers with skip connections, also known as residual layers. Skip connections bypass other layers, as shown in Figure 3.10 [HZR+15]. The motivation is that rather than learning the direct mapping from x to $\mathcal{H}(x)$ it is easier to learn $\mathcal{F}(x)$, which is the difference or the *residual* between x and $\mathcal{H}(x)$. Then $\mathcal{H}(x)$ can be computed by adding this residual. Using residual layers, together with batch normalization, allows the training of overly deep models with over 1000 layers. The gradient can backpropagate via shortcut connections; thus, mitigating the vanishing gradient problem introduced in Section 2.1. Deep ResNets use a bottleneck unit to reduce the number of computations, as shown on the right of Figure 3.10.

DenseNet connects each layer to every other layer [HLv+16]. Each layer's inputs are the concatenation of all feature maps from all the previous layers, which have a large memory footprint. On the flip side, DenseNet requires fewer weights than other similarly performing models.

Extreme Inception (Xception) combines design principles from VGG, Inception, and ResNet, and introduces *depthwise separable convolutions*, shown in Figure 3.11. In depthwise separable convolutions, the cross-channel correlations and spatial correlations are mapped separately [Cho16]. That is, every input channel is convolved with a different filter and the results are aggregated using a 1×1 filter called a *pointwise convolution*.

MobileNet, **MobileNet-v2**, and **MobileNet-v3** target hardware with limited power, compute, and memory, such as mobile phones. These models use depthwise separable convolution blocks with no pooling layers in between. MobileNet-v2 uses residual connections and adds

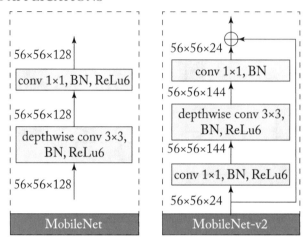

Figure 3.12: The MobileNet modules with arbitrary input tensor sizes using stride 1 for (left) v1 and (right) v2.

a channel expansion convolution layer prior to the depthwise separable convolutions, as shown in Figure 3.12 for stride 1 [HZC+17]. MobileNet-v3 uses AutoML, a technique discussed in Section 10.1. These models are served not just in mobile phones but also in data centers.

ResNeXt reintroduced group convolutions (initially used by AlexNet to distribute the model into two GPUs) [XGD+17]. In group convolution, the filters separate into groups, and each group operates on specific channels of the input tensor. The group convolution tensors are typically represented as a 5D tensor with the group id as the additional dimension. Depthwise separable convolution is a particular case of group convolution where the number of groups equals the number of channels of the input tensor.

ResNeXt replaces residual convolution blocks with residual group convolutions, shown in Figure 3.13, and every path of the group contains the same topology. These convolutions facilitate training and serving across multiple devices since each convolution in the group can be done independently of the other ones. ResNeXt is or has been used at Facebook [PNB+18].

NAS is a family of algorithms that learn both the topology and the weights targeting a particular hardware target, such as NASNet and EfficientNet [TL19]. EfficientNet was initially used on TPUs, but can be used with other hardware. Given their long training times and the diverse hardware fleet in data centers (multiple generations of CPUs and GPUs), the adoption of NAS-based models in the industry is still limited.

3.2.2   OBJECT DETECTION

Object detection involves finding or localizing objects of interest in an image and assigning a class label to them. Traditionally, object detection required a two-step approach: a region pro-

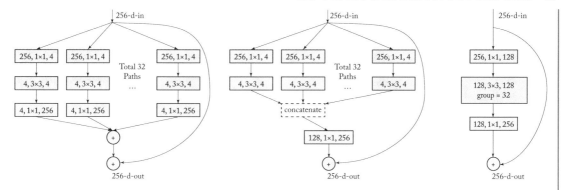

Figure 3.13: ResNeXt module with equivalent representations. ResNeXt uses residual group convolutions which are easier to parallelize across compute units. Based on [XGD+17].

posal step and a classification step. The input image is scaled up and down, known as an *image pyramid*, to detect objects of various sizes. New NNs can do these steps simultaneously, starting with the widely adopted Single-Shot Detector (SSD) and You Only Look Once (YOLO) models. Despite being relatively old, these models are still used in production because of their plug-and-play nature, where the object detector can use the latest CNN classifier as the base network.

Object detection models use a unified weighted cost function that accounts for the localization and the classification tasks. Also, object detectors generate several bounding boxes for a given object, and remove most of them using non-maximum suppression (NMS).

The most common metric to measure the detection accuracy is the mean Average Precision (mAP). The average precision (AP) is the area under the precision-recall curve and ranges from 0 to 1, with 1 being perfect detection for one class. The mAP is the mean AP across all the classes.

Key neural object detectors include Faster-RCNN, YOLO, SSD, RetinaNet, and EfficientDet.

Faster-RCNN uses a two-step approach with a region proposal network (RPN) and a classification network [RHG+15]. In Faster-RCNN, these two networks share a base CNN network or backbone architecture, which reduces the number of redundant computations. The base CNN model extracts feature maps from the image, which are passed to the RPN to generate and refine candidate bounding boxes, as shown in Figure 3.14. All the bounding boxes are then reshaped to be the same size and passed to the classifier. The Feature Pyramid Network (FPN) improved this topology; the predictions happen on high- and low-resolution feature maps [LGH+16].

YOLO divides the image into a 7×7 grid [RDG+16]. Each grid cell is responsible for 2 bounding boxes. Each bounding box is composed of the (x, y) center coordinate of an object, and

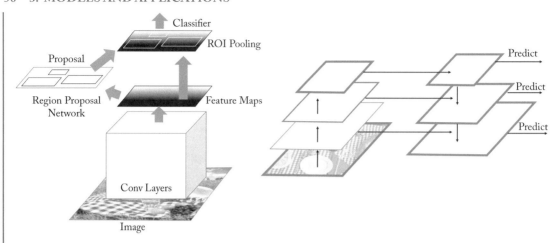

Figure 3.14: (left) The Faster-RCNN topology generates regions of interest in a feature map and jointly processes them. (right) The Feature Pyramid Network (FPN) topology can be used with the Faster-RCNN topology or other models to better detect objects at different scales.

the width, height, and confidence. That is, each bounding box has 5 values. The output of each grid cell is 5 values times 2 bounding boxes plus the probability of each class label given the input. If there are 20 classes, then each grid cell has an output vector of $5 \times 2 + 20 = 30$, and given the 7×7 cells, then the total number of output values for an input image is $7 \times 7 \times 30 = 1470$, as shown in Figure 3.15. In practice, the number of grid cells and bounding boxes are hyperparameters. The input image maps to the output via a CNN pretrained on an image classification task, such as the ImageNet dataset. YOLOv2 and YOLOv3 improves by detecting at three scales, using a deeper CNN topology, and having a class score for each bounding box [RF18].

Single-shot detector (SSD) uses an image classification model, such as VGG or MobileNet, as the base network and appends additional layers to the model [LAE+15]. Bounding boxes start from predefined anchor boxes. In each of the appended layers, the model refines or predict the bounding box coordinates, each with a respective score. Most of the computations are in the base network.

RetinaNet is the first one-stage detector model that outperforms the two-stage detection approach. The primary reason for previous one-stage detectors trailing in accuracy is the extreme class imbalance (many more background class samples). RetinaNet uses the *focal loss* function to mitigate this class imbalance [LGG+17]. The focal loss reduces the loss to well-classified examples.

EfficientDet is a scalable family of detectors based on EfficientNet. It uses a pyramid network for multiscale detection [TPL19].

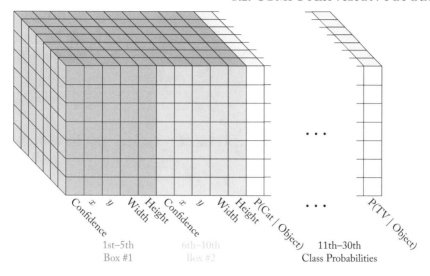

Figure 3.15: A YOLO model can map an input image to a $7 \times 7 \times 30$ grid output. Based on [Tsa18].

3.2.3   SEGMENTATION

Segmentation is a generalized and more challenging form of object detection, where every pixel in an image has a corresponding class label. Widely adopted models in the industry include Mask-RCNN and DeepLabv3 and in biomedical applications: U-Net, 3D-UNet, and V-Net.

Mask R-CNN extends Faster-RCNN by adding a separate output branch to predict the masks for all the classes [HGD+17]. This branch is in parallel to the bounding box predictor branch. Similar to Faster-RCNN, the choice for the base network is flexible.

DeepLabv3 uses *atrous convolution*, also known as dilated convolution, hole algorithm, or up-conv to increase the size of the feature maps by upsampling the weight filter, that is, inserting one or more zeros between each weight in the filters [CPS+17]. Atrous convolution, combined with Spatial Pyramid Pooling (SPP), is known as Atrous SPP (ASPP) and shown in Figure 3.16. ASPP can account for different object scales in the image.

U-Net is an encoder-decoder CNN [RFB15]. It uses convolutions to reduce the size of the receptive field, followed by transposed convolutions (or upsampling) to increase the size. U-Net also has skip connections between mirrored layers in the encoder and decoder stacks. This type of model is known as a fully-convolutional network (FCN) [LSD14]. U-Net can be trained with few images using data augmentation with multiple shifts and rotations.

3D U-Net and **V-Net** are 3D convolutional networks designed for voxel (3D pixels) segmentation from volumetric data [CAL+16, MNA16]. These models generally required the immense memory only available on server CPUs for training due to the large activations. Model

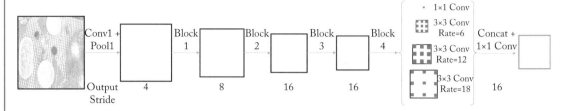

Figure 3.16: Atrous or dilated convolutions can maintain or increase the size of the feature maps. Based on [CPS+17].

parallelism techniques (discussed in Section 5.2) can be applied to train on GPUs and accelerators.

Detectron is a popular open-source platform developed by Facebook [WKM+19]. Detectron is implemented in PyTorch and contains implementations of various object detection and segmentation algorithms, which facilitates community adoption [WKM+19].

3.2.4   VERIFICATION

The task of verification is to determine whether a sample belongs to a particular set. The set size may be one, for instance, in the set of people with access to a personal bank account, or many, for instance, in the set of people with access to a building. Siamese networks are designed for this task.

Siamese networks learn a similarity function between two input images [BGL+93]. They can be trained by comparing an anchor image to a positive and negative image or, more precisely, comparing a metric of the distance between the feature vectors extracted from the images. The objective is to simultaneously minimize the distance between the anchor and positive image features and maximize the distance between the anchor and negative image features.

While Siamese networks are decades old, they can use modern techniques to improve their performance. For instance, CNN models can be a component of a Siamese network. The CNN models are trained across a variety of image appearances and used to extract features from the images [ZK15].

3.2.5   IMAGE GENERATION

The task of image generation requires modeling the distribution of images for a relative domain, such as written digits (the ultimate aspiration is modeling the distribution of all the natural images). Image generation is primarily an unsupervised learning task used in academia with some industry adoption for image deblurring, compression, and completion. The main types of algorithms used for image generation include auto-regressive and GAN models, specifically, PixelRNN, PixelCNN, DCGAN, 3D GAN, StackedGAN, StarGAN, SyleGAN, and Pix2pix.

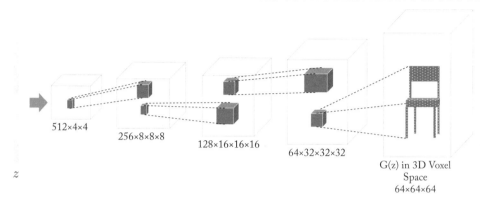

z

512×4×4

256×8×8×8

128×16×16×16

64×32×32×32

G(z) in 3D Voxel
Space
64×64×64

Figure 3.17: A 3D-GAN generator takes a random vector **z** and generates a 3D image. Based on [WZX+16].

PixelRNN and **PixelCNN** are auto-regressive models that predict the pixels along both axes using recurrent and convolutional layers, respectively. These models generate a conditional distribution over the 256 possible values for each RGB image channel at each pixel location [vKK+16].

DCGAN and **3D GAN** combine CNNs and GANs to generate 3D objects, as shown in Figure 3.17 [RMC15, WZX+16]. These GANs learn to generate high-quality objects by sampling from a low-dimensional space and passing those samples to the generator. **Stacked GANs** trains across multiple stacks of GANs, which results in higher quality image generation [HLP+17].

StarGAN and **StyleGAN** generate photorealistic images. For instance, they can generate human faces adjusting latent factors, such as freckles, hair color, gender, eyeglasses, and facial shape, when trained on a face dataset [CCK+17, KLA19].

Pix2pix is an adversarial network that learns a mapping from an input image to an output image and also learns a cost function to train this mapping. It can generate realistic images from labeled maps, colorize gray images, fill gaps in images, remove image backgrounds, and generate images from sketches [IZZ+16].

Other computer vision topologies that have been influential in the field, but do not currently have extensive commercial adoption are FaceNet for face recognition and verification, SqueezeNet and ShuffleNet for image classification on edge devices, SENet for high accuracy image classification, SRGAN for image super-resolution, and SqueezeSegV2 for road-object segmentation from LiDAR point cloud [SKP15, HSW+18, IHM+16, ZZL+17, HSA+19, LTH+16, WZZ+19]. OpenPose is used for pose estimation and has some adoption; Wrnch.AI uses a modified proprietary model to detect kinematics from 2D video.

3.3   NATURAL LANGUAGE PROCESSING TOPOLOGIES

NLP has been considered an AI-complete problem (requiring human-level intelligence to solve) given the complexity required to understand language. NLP is a required step toward automatic reasoning, that is, using stored knowledge to generate additional knowledge [Yam12]. Academia and industry have made tremendous progress in recent years.

Traditional NLP systems often use a hidden Markov model (HMM) (do not worry if you are not familiar with HMM). An HMM requires language experts to encode grammatical and semantic rules, provide a detailed ontology, parse the data, tag words with the appropriate part-of-speech, and iteratively align inputs and outputs. Neural NLP models can learn a particular task using a lexicon or language vocabulary and a massive data corpus and without explicitly programming the language rules. A popular benchmark to assess the performance of NLP models is the General Language Understanding Evaluation (GLUE) benchmark [SMH+18].

Hyperscalers use NLP algorithms for NLU, speech recognition, speech generation, and speech-to-speech translation tasks. NLU tasks include language translation, sentiment analysis, automatic document summarization, image captioning, document clustering, and question & answering. Speech recognition and speech synthesis are used as part of an NLP system by AI assistants, such as Apple Siri, Amazon Alexa, Google Assistant, Microsoft Cortana, and Baidu DuerOS. Speech-to-speech translation is used to interpret speech between different languages either as three separate stages (speech-to-text, text translation, and text-to-speech) or as a combined model. NLP algorithms facilitate human-machine interactions, enhancing a machine's ability to understand human language, and improve human-human communication, enabling communication between people without a common language.

3.3.1   NATURAL LANGUAGE UNDERSTANDING

NLU deals with machine reading comprehension. Neural machine translation (NMT) is the NLU task of mapping written text from a source to a target language using a NN. NMT topologies and techniques are also used for other NLU tasks with minor to no adaptations, such as sentiment analysis to categorize the sentiment of written product reviews, for question & answering systems, for document summarization, and for image captioning to "translate" a vector representation of the image into a description or caption.

Neural NLU models can be RNN-based, CNN-based, and transformer-based. They consist of an encoder that takes the source sentence and a decoder that outputs the target sentence. The decoder's task is to predict the next output word (or subword) given the previous outputs and the inputs. During the decode stage, a target output word can be greedily chosen from the softmax output or using a *beam search* approach as follows: The top n candidate words from the M softmax outputs at time t are selected and used as inputs into the next timestep. This results in nM output candidates at $t + 1$. The top n are selected from this nM group, and the process iteratively repeats in each subsequent timestep. At the last timestep, one output is selected

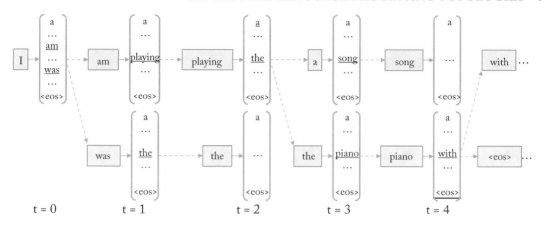

Figure 3.18: Beam search using a beam size of 2. Except for the initial decoder input at decoding timestep 0, every timestep uses the 2 most probable outputs (underlined) from the previous timesteps as inputs. At time $t = 4$, beam search results in the sentences: "I am playing the piano with ..." and "I am playing the piano <eos> ...," where <eos> is the end of sentence token.

from the nM choices. A common choice is $n = 10$ to provide a compromise between speed and accuracy. Figure 3.18 depicts a beam search of $n = 2$.

The quality of the target sentence in machine translation is typically reported using the BLEU score, a measure of similarity between the machine's translation and a professional human translation normalized by the sequence length. Other quality metrics have been proposed [NDC+17].

One implementation challenge during training is the variable sequence length. Using a constant sequence length to batch the sequences and either pad short sequences or truncate long sequences to the predetermined length so that each sample in a batch is of the same length mitigates this challenge.

The inputs to the NN are known as tokens. While earlier NLU topologies used words as tokens, most newer topologies use learned subwords [SHB15]. An algorithm segments words constrained to a fixed vocabulary size (the maximum number of subwords). These subwords are often interpretable, and the model can generalize to new words not seen during training using these subwords. Subwords are crucial for low-resource languages, that is, languages where the data corpus is small. The downside of using subwords rather than words is that a sequence has more tokens, which requires more computations.

Multi-language NMT involves learning a model used across multiple language pairs. They are particularly helpful for low-resource languages. Some care is needed to use them over simpler pairwise language models without sacrificing the performance of the translations from the high-resource language pairs [ABF+19]. Jointly learning the subwords across the combined languages

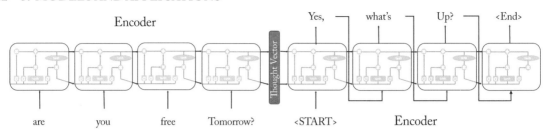

Figure 3.19: Encoder and decoder LSTM units for a question-and-answer system. The input sentence is represented by the *thought vector*.

has been shown to be beneficial [SHB15]. Google uses a multi-language NMT transformer-based model to support translation across 100 languages [ABF+19].

RNN-Based

During serving, RNN-based models are challenging to parallelize due to their sequential nature. A server CPU with fewer but more powerful cores than a GPU works well for RNN-based inference [ZRW+18]. These models are typically memory bandwidth bound, leaving much computational capacity unused. Some work demonstrates that their implementation can be modified to be more compute bound [Vil18]. ShaRNN provides an example of an RNN model with a small memory footprint, which is useful for edge deployments [DAM+19].

Despite the adoption of transformer-based models in commercial applications, RNN-based models continue to be used commercially due to their adequate statistical performance and low latency, and due to the larger memory and computational requirements of transformer-based models [Mer19].

Sequence-to-sequence (S2S) was the first widely adopted NMT model, and provides the foundation for similar models still used in production [SVL14]. The encoder LSTM units take as input (1) the state of the previous LSTM cell, (2) the output of the previous LSTM cell, and (3) the current token, as shown in Figure 3.19. The *thought vector* is the concatenated state vector and output vector of the last LSTM encoder unit. This thought vector is an encoding of the source sentence. The decoder takes the thought vector as input to the first decoder LSTM unit and produces a target word. Each subsequent unit takes the output from the previous unit as its input. This cycle continues until an LSTM outputs an end-of-sentence token. In practice, generating the target sentence in reverse order typically results in better quality translations.

Variants of the original S2S topology include models with multiple stacked bidirectional LSTM layers and bidirectional attention [SKF+16]. The term *NMT* is sometimes incorrectly used as a synonym for S2S or for GNMT.

Google's Neural Machine Translation (GNMT) is the most popular RNN-based model [WSC+16]. GNMT learns a better thought vector by simultaneously training across multiple languages and incorporates an attention module to cope with long sentences [LPM15]. The main idea of GNMT is that the thought vector should be the same regardless of the source and target language since it captures a meaning, which should be independent of the language.

CNN-Based

Using CNNs may have a computational advantage over RNNs, given they are more easily parallelizable and have a higher operational intensity (discussed in Section 7.3). Another advantage is they extract features hierarchically and may better capture complex relationships in the data.

Bai et al. demonstrated that CNN-based models outperform RNN-based models on various NLP long-sequence tasks [BKK18]. Similarly, Facebook demonstrated that CNN-based models had a computational advantage over GNMT at a similar statistical performance (both on CPUs and GPUs) [GAG+17]. When trained on models of the same size, the CNN-based models outperform GNMT.

CNN models have also been used as a preprocessing step to image captioning by extracting relevant features [VTB+14]. In particular, the second-to-last activation output in a CNN model is often used as the feature vector. This vector is an encoding of the image and passed to an NLU decoder to generate a caption. Attention can improve the captions by focusing the decoder on certain parts of the input image [YHG+15].

Transformer-Based

Transformer-based models use attention modules without any RNN units. The first transformer-based model, Transformer-LT, was introduced by Google in the 2017 paper *Attention is All You Need* and has been shown to statistically outperform RNN-based methods on various NLP tasks [VSP+17, KCH+19]. These models are more easily parallelizable than RNNs, can learn longer-term dependencies, and have higher arithmetic intensity.

A transformer primarily consists of a set of encoder and decoder blocks with the same structure but different weight values and with skip connections, as shown in Figure 3.20. Each encoder block consists of two main layers: a self-attention and a feedforward layer, where the self-attention block helps account for context in the input sentence. Each decoder block consists of three main layers: a self-attention, an encoder-decoder attention, and a feedforward layer. In the decoder, the encoder-decoder attention allows the decoder to focus on the crucial parts of the encoder representation. Words (or subwords, in practice) get embedded into vectors. A stack of encoders processes these vectors, and a stack of decoders processes their output. The architecture has skip-connections added and normalized after each layer. The target output word is chosen from the softmax output using a beam search approach.

Bidirectional Encoder Representations from Transformers (BERT) is a bidirectional transformer model developed by Google, and widely adopted across hyperscalers [DCL+18]. BERT achieved state-of-the-art results on multiple NLP tasks using a massive corpus of unannotated text crawled from the web, rather than a corpus labeled for a specific task. The standard

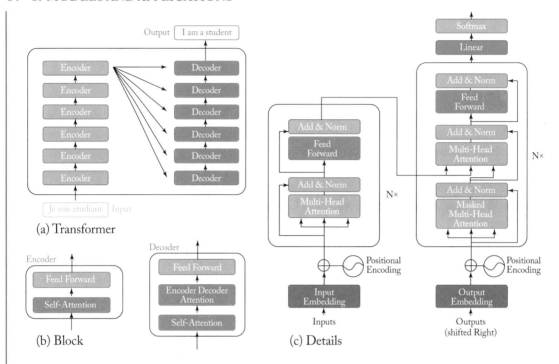

Figure 3.20: (a) A transformer is composed of several encoder and decoder blocks; (b) each block has an attention layer (the decoder has two) and a feedforward layer; and (c) the entire transformer model with $N\times$ blocks is depicted. Based on [Ala18, VSP+17].

embedding models before BERT, such as word2vec or GloVe (discussed in Section 2.7), learned context-free word embeddings, whereas BERT uses context to learn better embeddings. BERT is used by Google Search to better understand long search queries to improve the quality of the results [Nay19].

BERT is trained using two self-supervised learning tasks. In one task, the model predicts a randomly masked-out word based on the context of the words before and after it. In the second task, the model predicts whether the second sentence follows the first sentence in the original paragraph.

BERT and other transformer-based models are shown in Figure 3.21 and the most prominent are highlighted in Table 3.1. Typically, newer models better capture the dependencies between tokens [YDY+19].

Large transformer-based models require considerable power and compute to train and deploy. While hyperscalers widely use them, they are less common at companies without WSCs due to the training costs. Also, larger transformer-based models may not meet the stringent low latency inference requirements in some applications.

Table 3.1: Prominent transformer-based models

Model	Institution	Released	Parameters (millions)	Dataset (GB)
BERT [DCL+18]	Google	Oct. 2018	340	16
GPT-2 [RWC+19]	OpenAI	Feb. 2019	1,500	40
XLNet [YDY+19]	CMU	Jun. 2019	340	158
RoBERTa [LOG+19]	Facebook	Jul. 2019	355	160
ERNIE 2.0 [SWL+19]	Baidu	Jul. 2019	340	–
ALBERT [LCG+19]	Google	Sep. 2019	235	16
DistilBERT [SDC+19]	Hugging Face	Oct. 2019	66	16
T5 [RSR+19]	Google	Oct. 2019	11,000	750
Turing-NLG [Ros20]	Microsoft	Feb. 2020	17,000	–
GPT-3 [BMR+20]	OpenAI	May 2020	175,000	570
GShard [LLX+20]	Google	Jun. 2020	600,000	–

The Hugging Face Transformers, Facebook Fairseq, and AWS Sockeye 2 libraries contain several transformer-based models to facilitate wider adoption [DDV+20]. Future models are likely to compromise between prodigious costly models and smaller efficient models, trained and adopted by medium-size companies and universities, with smaller serving latencies. These include smaller BERT-like models, such as ALBERT by Google, DistilBERT by Hugging Face, and Q-BERT by UC Berkeley. Other solutions are replacing computationally expensive layers with light convolutions, adapting the number of attention layers, or removing most attention layers during inference to reduce serving latency [LCG+19, SDC+19, SDY+19, WFB+19, SGB+19, MLN19].

3.3.2   SPEECH RECOGNITION

Automatic speech recognition (ASR) is the task of converting acoustic sound waves into written text. ASR differs from voice recognition, where the task is to identify a person based on their voice. One of the main ASR challenges is the combinatory space of various aspects of speech, such as pace, accent, pitch, volume, and background noise. Also, serving an ASR model requires decoding acoustic signals in real-time with reliable accuracy. Neural ASR approaches have successfully overcome these challenges with large datasets without pronunciation models, HMMs, or other components of traditional ASR systems. Nassif et al. provide a systematic review of various neural ASR systems [NSA+19].

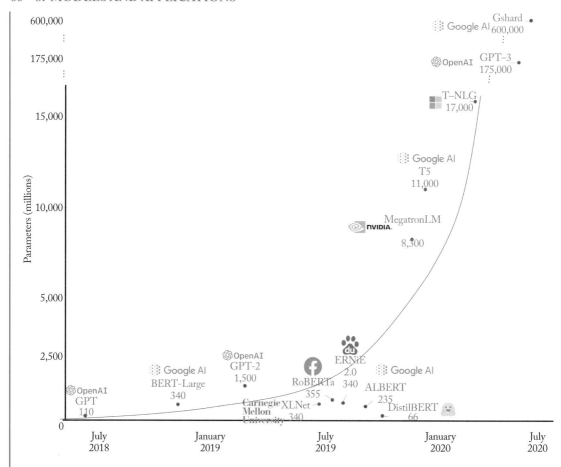

Figure 3.21: Multiple transformer-based models and their respective number of parameters across time. Based on [San19].

ASR systems and other speech-related systems often transform the acoustic sound waves into a spectrogram or Mel-spectrogram representation. A spectrogram is a 2D frequency-time representation of the acoustic signal that uses frequencies across short-time intervals, as shown in Figure 3.22. In the figure, the color represents the amplitude of a particular frequency at a specific time interval. The Mel-spectrogram is a spectrogram where the frequencies are scaled using the mel-scale to better match the frequency resolution of the human auditory system.

Deep Speech 2 (DS2) was developed by Baidu and is the first major neural ASR. It provides a baseline for other models. DS2 uses a spectrogram as the input to a series of CNN and RNN layers [AAB+15]. The CNN layers treat the spectrogram input as an image.

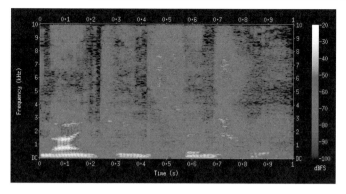

Figure 3.22: A spectrogram is a 2D frequency-time representation of the acoustic signal where the frequencies for short time intervals are captured. Source: [Wik18] (CC BY-SA 4.0).

Listen, Attend, and Spell (LAS) was developed by Google. This model uses SpecAugment for data augmentation. SpecAugment uses image augmentation techniques on the spectrogram [CJL+16, PCZ+19]. The LAS system has an encoder and decoder. The encoder is a pyramid RNN. The decoder is an attention-based RNN that emits each character conditioned on all previous characters and the entire acoustic sequence.

RNN-Transducer (RNN-T) processes the input samples and streams alphabetical character outputs. It does not use attention. For mobile devices, Google developed a quantized RNN-T model that runs in real-time on a Google Pixel device and is deployed with the Gboard app with 80 MB memory footprint [HSP+19, Sch19].

Wav2letter++ is an open-source neural ASR framework developed by Facebook; it uses the fully convolutional model ConvLM [PHX+18, ZXL+18]. Facebook also demonstrated the use of transformers for ASR [WML+19].

3.3.3   TEXT-TO-SPEECH

Text-to-speech (TTS) is the task of synthesizing speech from text. The most well-known TTS system is probably the one used by the late Prof. Stephen Hawking. A TTS system is typically composed of three stages: (1) a text-analysis model, (2) an acoustic model, and (3) an audio synthesis module known as a vocoder. Traditionally, audio synthesis modules combined short-speech fragments collected from a user to form complete utterances. Using these fragments makes it difficult to modify the tone or voice characteristics and results in a robotic-like synthesis.

Neural TTS systems are now able to generate human-like speech as measured by the MOS (Mean Opinion Score), a human evaluation of the quality of voice. A neural TTS model can learn to generate voices with different characteristics. They can also be adapted to generate music and speech from an image. Facebook uses automatic captioning to help visually impaired users browse their News Feed and hear a machine-generated caption of each image [WWF+17].

Google Duplex uses neural TTS models on Pixel phones, for example, to contact restaurants to make reservations [LM18].

The primary neural speech synthesis systems deployed in production are WaveNet, Parallel WaveNet, and WaveRNN and require a text-to-linguistic features preprocessing step. Tacotron 2 provides a full end-to-end text-to-speech generator. Deep Voice 3 and ClariNet are speech synthesizers (not end-to-end TTS) developed by Baidu that have been influential and may be used in production. GAN-based TTS is starting to gain traction in academia despite the earlier unknowns of how to use GANs with discrete values [BDD+20].

WaveNet by Google is a vocoder autoregressive model based on the PixelCNN model [vDZ+16]. It predicts a distribution for each audio sample conditioned on all previous audio samples and the input linguistic features. These features are derived from the input text and contain phoneme, syllable, and word information. To deal with long-range temporal dependencies needed for raw audio generation, WaveNet uses a stack of dilated causal convolutions to allow their receptive fields to grow exponentially with depth.

WaveNet suffers from high serving latency due to the sequential generation of audio samples. WaveNet uses an 8-bit integer value timestep (rather than a 16-bit, as is typical in audio) to reduce the latency and make the softmax output more tractable.

Parallel WaveNet by Google uses knowledge distillation to train a feedforward network with WaveNet [vLB+17, HVD15]. Knowledge distillation (detailed in Section 6.4) uses a teacher model to train a smaller, more efficient student model. The FFNN is easily parallelizable and generates speech samples in real-time with minimal accuracy loss compared to WaveNet. Google Assistant uses Parallel WaveNet.

Tacotron 2 by Google is a generative end-to-end model trained with audio and text pairs that synthesizes speech directly from characters and combines the methodologies of the popular WaveNet and Tacotron to generate human-like speech [SPW+17, WSS+17]. Specifically, Tacotron 2 uses CNN and LSTM layers to encode character embeddings into Mel-spectrograms, capturing audio with various intonations. This Mel-spectrogram is then converted to waveforms using a WaveNet model as a vocoder. This system can be adapted to generate speech audio in the voice of different speakers [JZW+18]. A speaker encoder network can generate a vector representation for a given speaker using seconds of reference speech from a target speaker. The Tacotron 2 network is adapted to generate speech conditioned on this vector representation.

WaveRNN by Google uses a dual softmax layer to predict 16-bit audio samples efficiently; each softmax layer predicts 8 bits. For real-time inference in mobile CPUs, the small model weights are pruned (removed or forced to zero) [KES+18]. LPCNet is a WaveRNN variant that achieves higher quality by combining linear prediction with the RNN [VS19].

Deep Voice 3 (DV3) by Baidu is is a generative end-to-end model synthesizer, similar to Tacotron 2 [PPG+17]. The primary difference is that Tacotron 2 uses a fully convolutional

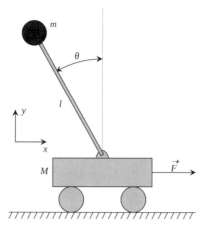

Figure 3.23: Reinforcement learning can be used to learn to balance the pole. Source: [Wik12] (CC BY-SA 1.0).

topology to map the character embeddings to a Mel-spectrogram. This convolutional topology improves the computational efficiency and reduces the training time.

ClariNet by Baidu extends DV3 as a text-to-wave topology and uses a similar WaveNet distillation approach [PPC18].

3.3.4   SPEECH-TO-SPEECH TRANSLATION

Speech-to-speech translation is the task of generating interpreted speech between language pairs. This task can be done in three separate stages: ASR, MT, and TTS. Combining some or all of these stages avoids compounding errors and may result in lowered inference latency, but it is challenging due to the limited data. Google developed a data augmentation process to improve the performance of a speech-to-translated-text (ST) system [JJM+18]. Google later developed Translatotron, an end-to-end direct speech-to-speech translation atttention-based sequence-to-sequence model [JWB+19].

In the near future, persons without a common language may communicate in real-time using neural speech-to-speech interpreters. The generated voice may maintain the same voice characteristics as the input voice or some other choice of voice [JZW+18].

3.4   REINFORCEMENT LEARNING ALGORITHMS

RL is used to teach an agent to perform certain actions based on rewards. The agent's goal is to take the required actions that maximize the cumulative reward over time. A simple task is the cart-pole balancing, depicted in Figure 3.23, where the reward depends on the height of the pole. The agent controls the cart and learns to balance the pole to maximize the reward.

The primary academic use of RL is in gaming, where the monetary cost of a mistake is minimal to none. RL beats human experts in Atari games, Doom, Super Smash Bros, Starcraft 2, Dota, Dota 2, chess, and Go [MKS+15, FWT11, VBC+19]. The OpenAI Five team defeated the professional human champions at DOTA 2 and demonstrated the use of RL for complex tasks, team cooperation, and strategizing in simulated environments [BBC+19].

In production, Covariant uses RL in robots to accelerate warehouse operations; it is one of the few examples of successfully using RL commercially with physical agents. JPMorgan's internal RL system LOXM is used to train trading agents [Mos17, BGJ+18]. Facebook uses the open-source ReAgent (formerly called RL Horizon) platform for personalized notifications and recommendations [GCL+19]. Microsoft acquired the Bonsai platform, designed to build autonomous industrial systems. Intel developed the Coach platform, which supports multiple RL algorithms and integrated environments, and is integrated into Amazon Sage-Maker RL. DeepMind built the TRFL platform and Google built the Dopamine platform (both on top of TensorFlow), and UC Berkeley released Ray with the RLlib reinforcement library to accelerate RL research [CMG+18, MNW+18]. A comparison of various platforms is found elsewhere [Win20]. Hardware targeting deep RL has been developed for edge applications [KKS+19].

It is impractical to train a physical agent in the physical world by allowing it to commit millions of errors. Rather, physics simulation engines simulate a real-world environment to train an agent. These simulators include DeepMind Control Suite environments, MuJoCo locomotion environments, and OpenAI Gym, which standardized the simulation environment APIs and had a significant influence on the field [TYD+18]. Other dynamic simulations include Bullet, Havoc, ODE, FleX, and PhysX [ETT15]. However, more realistic simulators are needed to transfer the learnings to physical agents. These agents are then fine-tuned in the physical world. Alternatively, interleaving simulation with some real-world rollouts works for simple tasks [CHM+19].

Other challenges with RL are debugging and reward selection. For some tasks, care is needed to ensure the reward is aligned with the programmer's end goal for the agent [AC16, BEP+18]. RL can be difficult to debug because the lack of learning may be due to many factors, such as a suboptimal reward, a large exploration space with sparse rewards, or an issue with code. As a general guideline, it is best to start with a simple algorithm and incrementally increase the complexity. Simpler Monte Carlo Tree Search or Genetic Algorithms can tackle simple tasks.

RL algorithms often run multiple agents on CPUs; one per core [CLN+17]. Recent work, such as the OpenAI Rapid system, shows that leveraging both CPUs and GPUs can improve performance [SA19].

The three families of RL algorithms, shown in Figure 3.24, are Q-learning, policy optimization, and model-based.

Q-learning, also known as **value-based**, learns the quality of the agent's state and action. DeepMind popularized Q-learning in 2013 with the Deep Q-network (DQN) algorithm

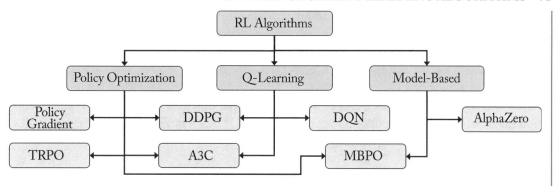

Figure 3.24: Families of deep RL algorithms. Based on [Ope18].

showing superior performance than previous RL methods across various Atari games (soon after, Google acquired DeepMind, and is now a sibling company to Google under Alphabet) [KSe+13]. Using a variety of Q-learning models achieves better performance over any single Q-learning model [HMv+17].

Policy optimization, also known as **on-policy**, learns the policy function and selects the output action stochastically. A policy is the agent's strategy. A policy function maps the input state to a distribution of actions, and a DL model can represent this function.

Policy optimization was popularized by the Policy Gradient (PG) algorithm that showed superior performance over DQN [MKS+15]. The space is explored initially through random actions. Actions that lead to a positive reward are more likely to be retaken.

A primary challenge is the sparse delayed rewards, formally known as the credit assignment problem. The agent receives a reward after taking several actions. The reward can be positive or negative. Depending on the reward, all the actions taken are considered good or bad, even if only some of them were critical to receiving the reward. Given the sparse rewards, policy optimization requires lots of training samples. Alternatively, manually shaping the rewards for a pasticular task can guide the learning behavior. Trust Region Policy Optimization (TRPO) is typically used over vanilla PG as it guarantees monotonic policy improvements [SLM+17]. A comparison of TRPO to DDPG and other PG-based algorithms, such as Proximal Policy Optimization (PPO) and Actor-Critic using Kronecker-Factored Trust Region (ACKTR) can be found elsewhere [HIB+19].

Various algorithms combine Q-learning and policy optimization methodologies. The most popular ones are A3C and DDPG [MBM+16, LHP+19]. Asynchronous Actor-Critic Agents (A3C) uses a policy-based actor and a value-based critic to measure how good is the chosen action. Deep Deterministic Policy Gradients (DDPG) uses continuous (rather than discrete) actions. While TRPO, DDPG, and A3C are typically good algorithms to use, experimentation is required to determine the most suitable for a particular task.

Model-based algorithms use a model with the rules of their environment. The agent uses the model to infer the outcomes of various sets of actions and chooses the set with the maximum reward. Model-based algorithms are used in games like chess and Go, where the rules of the game are known. DeepMind's AlphaGo, AlphaGo Zero, AlphaZero, and MuZero use model-based algorithms [SHM+16, SSS+17, SSS+18, SAH+20]. Learning a model through trial and error introduces biases, and errors in the inferred outcome compound over the prediction horizon. Model-based policy optimization (MBPO) uses a model with policy optimization to mitigate the compounding errors [JFZ+19].

In this chapter, we detailed the types of workloads that typically use DL models at hyperscalers: recommenders, computer vision, and NLP. We discussed the common topologies used in each of these workloads. Despite having the smallest adoption in academia, top hyperscalers widely use recommender models. We highlighted popular academic trends in RL that may soon transition to commercial applications. In the next chapter, we review how to train a topology, including how a data scientist may use an existing topology to guide the topology design for a related application.

CHAPTER 4

Training a Model

Training a model to achieve high statistical performance within a computational and power budget requires several design considerations. These include defining a topology, preparing the dataset, properly initializing the model weights, selecting an optimization algorithm and objective function, reducing the model size, and evaluating the trained model. The training process can be computational and memory intensive, and there are techniques discussed in this and the next two chapters to reduce the training time and mitigate memory bottlenecks.

In Section 1.6, we introduced the training steps. The training stops when the validation error is either less than some threshold or does not continue to decrease after several iterations. The validation error is computed every n training iterations, where n is chosen by the data scientist. It is used as a metric of how the model will perform when it is deployed.

During the backpropagation step, the computed gradients provide a measurement of the contribution of each weight to the cost. The terms cost, loss, penalty, error, and objective function, are sometimes used interchangeably. In this book, *loss* represents a metric of difference between the expected output and actual output for one data sample, and *cost*, error, and objective function synonymously represent the sum of the losses for a batch of samples. Examples of common objective functions are the cross-entropy error (discussed in Section 4.4) and the mean square error (MSE) for classification and regression tasks, respectively.

In the remainder of this chapter, we detail how to train a model to achieve low training and low test error. We review techniques to improve the performance on each of the training steps outlined in Section 1.6. We provide the methodologies that experienced data scientists use in industry to deal with unbalanced datasets, design new topologies, resolve training bugs, and leverage existing pre-trained models. We also discuss methods to reduce memory bottlenecks. Distributed training algorithms to reduce the training time are discussed in Chapter 6. A review of the notation introduced in Section 1.10 can help understand the equations presented in this chapter.

4.1 GENERALIZING FROM TRAINING TO PRODUCTION DATASETS

A well-designed and trained model has good performance on production data not used during training. That is, the model generalizes from the training dataset to the production or test dataset. Specifically, the model has low error rates in both the training and test datasets. On the contrary, a model with high test error rates is unreliable. In this section, we describe the

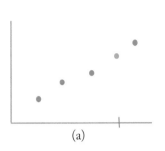
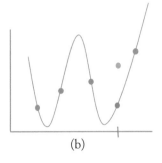
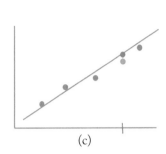

(a)   (b)   (c)

Figure 4.1: (a) Four training samples (blue dots) and one validation sample (green dot). (b) A fourth-order polynomial function has zero training error but high validation error. (c) A simpler first-order polynomial function has low validation error. The red dot represents the model's prediction.

source of high test error rates, specifically, underfitting, overfitting, and sharp minima, and how to reduce this error. The red dot represents the model's prediction.

Underfitting occurs when the model is too small because it has too little learning capacity and cannot properly learn the general characteristics of the data. The symptoms of underfitting are high training error and high test error. The best technique to mitigate underfitting is to use a more complex model. In DL, this means increasing the topology's representative capacity by adding more layers and more weights.

Overfitting occurs when a model has too much learning capacity and learns to fit the noise in the training data samples or other characteristics unique to the training set. Overfitting happens when using a prodigious model with insufficient training samples. The symptoms of overfitting are low training error and high test error. Figure 4.1 illustrates overfitting with a toy 1D example using linear regression, a simple ML algorithm. Figure 4.1a shows four training samples (the blue dots) and one validation sample (the green dot) not used during training. The x-axis is one feature, such as house size, and the y-axis is the label, such as house price. A polynomial function of third or higher-order can perfectly pass through the four training data points. The illustration uses a fourth-order polynomial for simple visualization. Figure 4.1b shows the model has no training error but has a higher validation error (the squared distance between the red and green dots). A simpler first-order (affine) function does not perfectly pass through all the training data points but has low validation error, as shown in Figure 4.1c. The red dot shows what each model predicts on the validation sample, and the green dot is the ground truth for that sample. The complex model overfits the training samples; it has zero training error but high validation error compared to the simpler model. Therefore, in this example, the simpler model is preferred.

Figure 4.2 illustrates what happens to the training and validation error as the model grows in complexity. While the training error decreases with more complexity, the validation error first

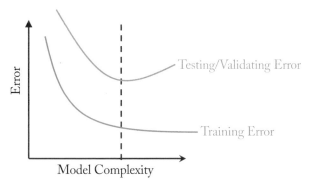

Figure 4.2: The ideal level of model complexity is where the validation error is the lowest.

decreases and then increases. A model with complexity left of the dashed line is underfitting, and a model with complexity right of the dashed line is overfitting. The sweet spot is right at the dashed line, where the model has the lowest validation error. The model is complex enough to learn the characteristics of the data to avoid underfitting but simple enough to avoid overfitting.

The validation error is much more important than the training error because it represents the expected error when the model deploys in production. In ML theory, minimizing these errors is known as the bias-variance tradeoff. A high training error indicates high bias or underfitting. A high validation error and low training error indicates high variance or overfitting. It is always critical to determine the source of poor performance (overfitting or underfitting) before prescribing a solution.

An interesting and counterintuitive phenomenon unique to various DL topologies is the deep double descent, illustrated in Figure 4.3 [NKB+20]. As the topology complexity increases (that is, as the model grows in depth), the validation error first follows the expected trajectory of decreasing and then increasing, but then it begins to decrease again. That is, increasing the size of the topology can lower the test error in some scenarios. The exact reason is not well understood as complex models should result in overfitting. A tentative (hand-wavy) reason is that very large topologies can explore a larger solution space leading to superior solutions. Understanding this phenomenon and the impact on the recommended training techniques is ongoing research. Most practitioners safely ignore this phenomenon or are not aware of it.

Another source of poor generalization may be sharp minima [HS97]. This hypothesis is based on empirical evidence. Figure 4.4 illustrates the intuition with a toy 1D example using only one weight or feature (the x-axis). Training involves iteratively updating the model and moving to an area in the solution space with lower training error. The training cost function (solid blue line) is similar but slightly different than the testing cost function (dotted green line). This difference is because the test samples are similar but not identical to the training samples. In this example, the flat minimum solution and the sharp minimum solution have the same

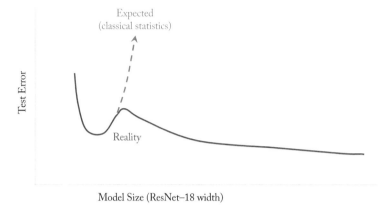

Figure 4.3: An illustration of the deep double descent observed in some DL topologies; as the complexity increases, the validation error decreases and then increases as expected, but then it begins to decrease again. Based on [NKB+20].

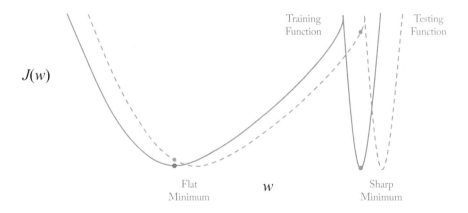

Figure 4.4: In this toy example, the cost function with respect to the test dataset is slightly shifted from the cost function with respect to the training dataset. The sharp minimum solution has a high test error. The flat minimum has a small test error. Based on [KMN+17].

training error but different test errors. These errors are represented by $J(w)$ along the y-axis. The flat minimum solution has a low test error, while the sharp minimum solution has a high test error (the green dot). A measurement of flatness is the trace of the Hessian; a small trace indicates a flat minimum [DYC+19].

While a flat minimum generalizes better to unseen data, a sharp minimum does not necessarily indicate overfitting, and a flat minimum does not necessarily indicate low validation error [ML18]. Also, the functions resulting in a flat minimum can be altered to result in a sharp minimum without affecting the validation error, demonstrating the hypothesis above does not always hold [DPB+17].

There are various techniques to improve generalization, often by simplifying (regularizing) the model. The most common ones are as follows:

Larger datasets is the best technique to avoid overfitting. The toy example above only used four samples to train the fourth-order polynomial. Adding more samples while keeping the same model complexity (fourth-order polynomial) results in a more affine-like function that better generalizes to data not in the training set. OpenAI recommends for NLP models increasing the number of parameters by 2.55× whenever the dataset doubles to improve learning capacity and avoid over/underfitting [KMH+20].

Weight decay (also known as L_2-**regularization**) penalizes the magnitude of the weights and reduces overfitting. In the fourth-order polynomial example above, this would penalize the magnitude of the coefficients and result in a more affine-like function. The objective function incorporates the weight decay by adding a penalty term:

$$\text{new cost} = \text{cost} + \lambda ||\mathbf{w}||_2^2,$$

where $\lambda \geq 0$ is the regularization factor and $\mathbf{w}$ is the model weights (the polynomial coefficients in the regression example above). The bias weight does not have a multiplicative interaction with the activations; therefore, it is not regularized. Note that L_1 (rather than L_2, as shown above) regularization is less common.

Smaller batches improves generalization [ML18]. A training iteration involves processing a batch of data. Larger batches can have computational advantages (they have higher data reuse), but often large batches result in sharp minima. The ideal is a medium size batch where the model converges to a flat minimum and has high compute utilization. Finding an adequate batch size requires experimentation.

Better optimizer that finds a solution with a lower validation error. In Section 4.3, we discuss the gradient descent optimizer and others less prone to sharp minima solutions, such as LARS, LAMB, and RangerLARS.

Topology pruning means forcing some of the smaller weights to zero or removing parts of the model. In Section 6.3, we discuss pruning in more detail.

Label-smoothing regularization (LSR) modifies the ground-truth one-hot vector by adding a small ϵ/M value to all the zero entries, where M is the number of classes and ϵ is a small value, such as $\epsilon = 0.1$ [SVI+15]. The "1" entry in the one-hot vector is changed to $1 - \epsilon$ to maintain a valid probability distribution. Reducing the difference between the largest logit and all others reduces the confidence of a model and results in better adaptation to non-training samples.

Early stopping means the training stops when the validation error begins to increase. Similarly, the model is evaluated on the validation dataset and saved every n training iterations, and the model with the lowest validation error is selected. There are mixed opinions on using early stopping. Regularization via weight decay without using early stopping can lead to better results when the computational resources are available to experiment with multiple weight penalties. In practice, early stopping is a simple and effective technique to reduce overfitting and commonly used. Note, somewhat related, that Hoffer et al. demonstrated better generalization with additional training cycles when the validation error has plateaued, but the training error continues to decrease [HHS17].

Model ensemble is where an ensemble (group) of models is trained for a particular task. During inference, a combination of the models' predictions is used, such as the average. Combining the predictions reduces the impact of each model overfitting. More formally, model ensemble reduces the variance of the validation error.

In addition, normalization and dropout (discussed in Sections 2.6 and 2.9) are other forms of regularization which reduce overfitting.

4.2   WEIGHT INITIALIZATION

Training a model is the process of learning the weight values for a topology for a particular task. The initialization of the model weights at the start of training can significantly impact the learning (training convergence), particularly for deeper networks.

Initializing all the weights to the same value results in the weights having the same update and therefore prevents learning. The weights (including the biases) are typically sampled from random distributions. They are initialized such that the distribution of the *activations* has unit variance across a layer. This initialization reduces the likelihood of having exploding or diminishing gradients during the backpropagation step when multiplying gradients across several layers.

A simple initialization approach is to sample from a zero-mean normal (Gaussian) distribution or from a uniform distribution using a different standard deviation for each layer. A common choice when the activation function is ReLU is the Kaiming initialization: the weights are sampled from a normal distribution with standard deviation $\sigma = \sqrt{2/D^{(l)}}$, where $D^{(l)}$ is the number of units in Layer l [HZR+15]. A truncated normal (the sides of the distribution are truncated) is recommended to prevent initializing the weights with large magnitudes. Kaiming initialization allows the training of much deeper networks. Before this technique was developed, the authors of the well-known VGG paper meticulously initialized the layers of the larger VGG networks in various steps. With Kaiming's initialization, this is no longer needed.

For sigmoid or hyperbolic tangent layers, the Xavier initialization is preferred [GB10]. The weights at Layer l are sampled from a uniform distribution $\mathcal{U}(-k, k)$ where

$$k = \sqrt{\frac{6}{D^{(l)} + D^{(l+1)}}}.$$

These initialization techniques can be adapted to train hypernetworks, meta-NNs that generate weights for a primary NN [CFL20].

Bias Initialization

It is common to initialize the bias weights to zero. Exceptions are as follows:

- The bias of the last layer in a model for binary classification trained with imbalanced datasets (far more negative than positive samples) should be initialized to [Kar19]

$$\log_e \frac{\text{number of positive samples}}{\text{number of negative samples}}.$$

- The bias of the last layer in a regression model trained with imbalanced datasets should be initialized to the expected mean output value. Alternatively, the data targets should be normalized, and the bias initialized to 0.

- The bias of the LSTM forget gate should be initialized to 1 to prevent the LSTM unit from forgetting at the start of training. The model needs some training cycles to learn to forget [GSC99, JZS15].

- The bias of the LSTM input and output gates should be initialized to -1 to push the initial memory cell activations toward zero [HS97].

- The bias in a ReLU layer may be initialized to a positive value to reduce the number of zero activations that may cause the dying ReLU phenomenon [Ste19]. However, the benefits have not been extensively explored.

4.3 OPTIMIZATION ALGORITHMS: MINIMIZING THE COST

In supervised DL, the input data is forward-propagated through the model, and the output is compared to the expected output (the ground truth) to compute a penalty or cost. For a given topology and dataset, there is a cost (an objective function) landscape, that is, a cost associated with all the possible weight values. The goal of training a topology is to find a set of weights (a model) that has a low cost.

Optimization algorithms iteratively update the weights to reduce the cost. A useful optimizer efficiently searches the high-dimensional solution space and converges to a low cost, flat minimum. In DL, the weight (parameter) space typically ranges from a few million to hundreds of billions of dimensions, and it has a roughly convex objective function with walls or barriers on the valley floor [XAT+18]. The valley floor has several local minima. A given topology converges to different local minima in each training run due to the stochasticity in training. Interestingly, different minima solutions typically have comparable statistical performance (cost).

The most common optimizers used in production are Stochastic Gradient Descent with Momentum (SGDM) and Adam, sometimes with a preference for computer vision and NLP models, respectively. Before introducing them, we first introduce gradient descent and stochastic gradient descent (SGD) to motivate the utility of SGDM and Adam. We also discuss LARS and LAMB, which ushered the use of large-batches. These optimizers use the gradients computed during the backpropagation step to update the model weights. In the next section, we detail how to compute the gradients. Orthogonal techniques, such as SWA and LookAhead described below, can be used in conjunction with the optimizer to find a better minimum.

In gradient descent (GD), also known as *steepest descent*, all the data samples in the dataset are used to compute the objective function. The weights are updated by moving in the direction opposite to the gradient, that is, moving toward the local minimum. The objective function $J(\mathbf{w})$ is computed using the sum of all the losses across the dataset with N samples. The set of weights are updated as follows:

$$J(\mathbf{w}) = \sum_{n=0}^{N-1} loss\left(f_{\mathbf{w}}\left(\mathbf{x}^{[n]}\right), \mathbf{y}^{[n]}\right)$$

$$\mathbf{g} = \frac{dJ(\mathbf{w})}{d\mathbf{w}} = \nabla_{\mathbf{w}} J(\mathbf{w})$$

$$\mathbf{w} := \mathbf{w} - \alpha \cdot \mathbf{g},$$

where $\mathbf{w}$ represents all the weights in the model and α is the learning rate (LR). Note that a weight decay term (see Section 4.1) is used in practice; it is excluded from all the equations in this section to simplify notation.

The LR controls the change of the model in response to the gradient and is the most critical hyperparameter to tune for numerical stability [Ben12]. In Section 4.5.4, we provide recommendations on tuning this and other hyperparameters. Figure 4.5 shows a GD update toy example in a 1D space using different LRs. A high LR can cause the model to diverge, where the cost increases rather than decreases. A small LR can result in longer-than-needed number of convergence steps and training time. A good LR results in proper progress toward the minimum (the green arrow in the figure).

In SGD or, more precisely, mini-batch gradient descent (MBGD), the dataset is divided into several batches. In statistics literature, SGD means MBGD with a batch size of 1, but in most DL literature and in this book, SGD refers to MBGD with any arbitrary batch size less than the training dataset. When the batch size equals the full-batch, SGD becomes GD, and one epoch equals one training iteration. In SGD, the gradient used to update the model is computed with respect to a mini-batch (as opposed to the entire dataset), as shown in Figure 4.6, and otherwise, the implementation of SGD and GD are equivalent.

There are two main challenges with GD and large-batch SGD. First, each step or iteration is computationally expensive as it requires computing the cost over a large number of samples. Second, the optimizer may converge to a sharp minimum solution (rather than stuck at a saddle point as previously thought) that often does not generalize, as shown in Figure 4.4 [ML18, YGL+18, DPG+14].

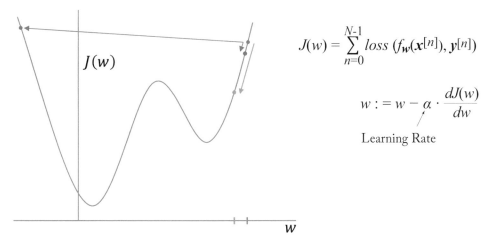

$$J(w) = \sum_{n=0}^{N-1} loss\ (f_{\boldsymbol{w}}(\boldsymbol{x}[n]), \boldsymbol{y}[n])$$

$$w := w - \alpha \cdot \frac{dJ(w)}{dw}$$

Learning Rate

Figure 4.5: Gradient descent update using LRs that are (red arrows) too large or too small, and (green arrow) good enough.

The Hessian (this is the second derivative in 1D) can be used to analyze the curvature of the objective function along the various dimensions to determine if a solution is in a flat or sharp minimum. Smaller absolute eigenvalues indicate a flatter curvature in the corresponding dimension, and the average Hessian trace provides a metric for the average curvature across all dimensions; a higher trace value indicates a sharp minimum [DYC+19].

The algorithmic reasons for the convergence to a sharp minimum are not well understood. One hypothesis is that the objective function has many sharp minima and gradient descent does not explore the optimization space but rather moves toward the local minimum directly underneath its starting position, which is typically a sharp minimum [KMN+17]. This hypothesis is at conflict with the hypothesis that the objective function is roughly convex [XAT+18]. Additional research is required to understand the reasons better.

The batch size is an important hyperparameter to tune. A larger batch size has higher compute utilization because there is more data reuse; that is, the compute-to-data-read ratio is higher for larger batches. However, using very large batches suffers from the same challenges as GD and requires meticulous tuning to avoid converging to a sharp minimum. Still, using a micro-batch is not ideal because the computational resources are tipically underutilized. Furthermore, micro-batches do not have sufficient statistics to properly use batch normalization [Iof17]. There is a sweet spot of a batch size where it is large enough to use the hardware compute units efficiently and small enough for the model to properly converge to a flat minimum without too much hyperparameter tuning.

Shallue et al. demonstrated empirically across several models and datasets, that for a given optimizer and a model, there are three batch size regions. There is a perfect scaling region, where

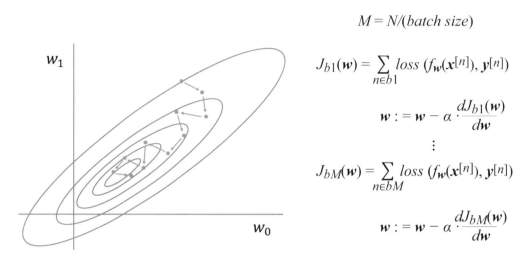

$$M = N/(batch\ size)$$

$$J_{b1}(w) = \sum_{n \in b1} loss\ (f_w(x^{[n]}),\ y^{[n]})$$

$$w := w - \alpha \cdot \frac{dJ_{b1}(w)}{dw}$$

$$\vdots$$

$$J_{bM}(w) = \sum_{n \in bM} loss\ (f_w(x^{[n]}),\ y^{[n]})$$

$$w := w - \alpha \cdot \frac{dJ_{bM}(w)}{dw}$$

Figure 4.6: The dataset is broken into M batches, and the weight vector (two dimensions in this toy example) is updated using the gradient computed with respect to the cost associated with a batch. The progress toward the minimum (the inner oval) is not smooth (unlike in GD) but faster than GD: for every 1 GD step, SGD takes M steps.

Table 4.1: Batch size scaling regions across the three models observed in Figure 4.7

Model and Dataset	Perfect	Diminishing	Stagnation
Simple CNN on MNIST	≤ 128	128–2,048	$\geq 2,048$
Transformer on LM1B	≤ 256	256–4,096	$\geq 4,096$
ResNet-50 on ImageNet	$\leq 8,192$	8,192–65,536	$\geq 65,536$

the batch size and LR proportionally increase and the number of training iterations proportionally decreases. There is a diminishing-returns region, where increasing the batch size decreases the number of iterations but not proportionally. And there is a stagnation region, where increasing the batch size provides minimal to no benefits. The stagnation occurs because the gradients computed with a large-batch have low variance. They already closely approximate the GD gradient, and increasing the batch size further does not result in significantly different gradients. Furthermore, as already discussed, very large batches may converge to sharp minima. Figure 4.7 captures some of their results on three popular models and datasets and Table 4.1 summarizes the results in the figure [SLA+19]. In Section 4.5.4, we discuss hyperparameter tuning, which includes choosing a batch size.

Training iterations should (on average) decrease the training error. A plateau training error indicates that the solution is bouncing along the edges of the objective function and no

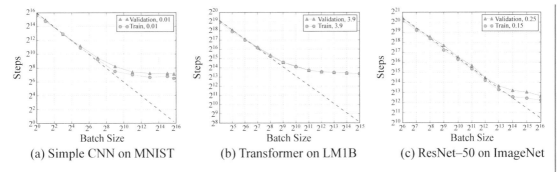

(a) Simple CNN on MNIST   (b) Transformer on LM1B   (c) ResNet–50 on ImageNet

Figure 4.7: The number of training steps required to meet the expected training and validation error as a function of batch size for three models. Dotted line denotes perfect scaling. See Table 4.1 for the high-level summary. Source: [SLA+19] (CC BY-SA 4.0).

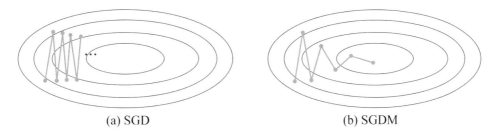

(a) SGD   (b) SGDM

Figure 4.8: Toy example of a 2D space with a ravine. (a) SGD makes slow progress. (b) SGDM makes faster progress toward the minimum. Based on [Orr99].

longer converging. Decreasing the LR can help the error continue to decrease and converge to a solution closer to the local minimum. A better approach may be to use a cyclical LR between a user-set high and low LR to better explore the solution space, in particular toward the later part of training [LH17, Smi17, IPG+19]. Each learning cycle starts at the high LR, which decreases with each iteration. After reaching the low LR, another learning cycle starts (at the high LR). This technique can be applied with all the optimizers.

SGDM improves the speed of convergence over SGD alone [Qia99]. Most training in the literature that claims SGD actually used SGDM. That is, the term *SGD* is often an alias for SGDM in published literature but not in this chapter to avoid confusion. SGD alone makes slow progress in ravines (areas where the partial derivative in one dimension is much higher than other dimensions), as shown in Figure 4.8. Ravines are prevalent when optimizing over millions of dimensions, which is common in DL models.

SGDM accelerates SGD in the direction of the exponential decaying average of past gradients, also known as the *first moment* or just *moment*, and dampens oscillations. Rather than directly modifying the weights, the gradients modify this moment, and the moment is then used to update the weights as follows:

$$g = \nabla_{\mathbf{w}} J(\mathbf{w})$$
$$\mathbf{m} := \beta \cdot \mathbf{m} + (1 - \beta) \cdot \mathbf{g}$$
$$\mathbf{w} := \mathbf{w} - \alpha \cdot \mathbf{m},$$

where $\mathbf{m}$ is the (exponential decaying) average gradient or first moment that gets decayed by the momentum term β usually set to $\beta = 0.9$, $\mathbf{m}$ is initialized to $\mathbf{m} = \mathbf{0}$, and α is the LR which requires tuning. SGDM is widely adopted in the industry, in particular, for computer vision models, and works well across multiple tasks when the learning rate is properly tuned.

Adaptive Moment Estimation (Adam) is more robust than momentum to different LRs, and therefore requires less LR tuning [KB17]. Adam computes an adaptive LR for each parameter. Specifically, Adam uses an average gradient (as in SGDM) normalized by an average gradient squared called the *second moment* or *variance*. Thus, every weight is updated with a different LR as follows:

$$g = \nabla_{\mathbf{w}} J(\mathbf{w})$$
$$\mathbf{m} := \beta_1 \cdot \mathbf{m} + (1 - \beta_1) \cdot \mathbf{g}$$
$$\mathbf{v} := \beta_2 \cdot \mathbf{v} + (1 - \beta_2) \cdot \mathbf{g}^2$$
$$\hat{\mathbf{m}} = \mathbf{m}/(1 - \beta_1^t)$$
$$\hat{\mathbf{v}} = \mathbf{v}/(1 - \beta_2^t)$$
$$\mathbf{r} = \hat{\mathbf{m}}/(\sqrt{\hat{\mathbf{v}}} + \epsilon)$$
$$\mathbf{w} := \mathbf{w} - \alpha \cdot \mathbf{r},$$

where $\mathbf{m}$ and $\mathbf{v}$ are the first and second moment estimates, $\hat{\mathbf{m}}$ and $\hat{\mathbf{v}}$ are the bias-corrected first and second moment estimates, respectively, $\mathbf{g}^2$ is the element-wise squared of $\mathbf{g}$, vector division is element-wise division $\mathbf{m}$ and $\mathbf{v}$ are both initialized to $\mathbf{0}$, $\beta_1 \in [0, 1)$, $\beta_2 \in [0, 1)$, and $\epsilon > 0$ are usually set to $\beta_1 = 0.9$, $\beta_2 = 0.999$, and $\epsilon = 0.001$, the exponent term t is the training iteration and α is the LR which requires some tuning.

Intuitively, a small variance in the gradients means the gradients are pointing in similar directions, which increases the confidence that the direction is right. Therefore, a larger step in that direction is taken using a larger LR. The opposite happens with a large variance: a small step is taken.

When switching from SGD to Adam, the regularization hyperparameter needs to be adjusted since Adam requires more regularization [LH19]. While the original paper used $\epsilon = 10^{-8}$, we recomend $\epsilon = 10^{-3}$ to prevent a huge step size when $\hat{\mathbf{v}}$ is miniscule, which often happens toward the end of training [KB17].

Adam is widely adopted in the industry, in particular, for NLP models, and empirically works well across multiple tasks despite not converging to the optimal solution in simpler convex optimization tasks [RKK19]. SGDM continues to perform well or better across various tasks when the LR is well tuned compared to newer techniques. SGDM often converges and generalizes better, albeit with longer training time, than Adam [WRS+18, KS17]. Some practitioners begin training with Adam due to the convergence speed and finish with SGDM due to the convergence quality.

Rectified Adam (RAdam) is a simple adaptation to Adam that switches between Adam and SGDM [LJH+19]. RAdam dynamically turns on or off the adaptive LR depending on the variance confidence. Thus, Adam's possible initial training instability due to the limited data points used to compute the variance is mitigated with this on/off adaptive LR. RAdam uses a rectified adaptive LR as it gains confidence about the variance; otherwise, it falls back to SGDM.

All the above optimizers share a common challenge that LARS and LAMB addresses. To maintain stability, weights with a small magnitude should have a small weight update magnitude, and vice versa. However, every layer in a model often has vastly different $\frac{||w^{(l)}||}{||g^{(l)}||}$ ratios. A small ratio can lead to training instability (divergence), and a large ratio can lead to slow learning. LARS and LAMB improve training stability by normalizing the step size in each layer. This additional stability allows training with large-batches (up to some size determined experimentally).

Layer-wise Adaptive Rate Scaling (LARS) uses a local LR $\alpha^{(l)}$ proportional to the ratio of the magnitude of the weights to the magnitude of the gradients [YGG17]. LARS is applied to SGD as follows:

$$\alpha^{(l)} = \frac{||\mathbf{w}^{(l)}||}{||\mathbf{g}^{(l)}||}$$
$$\mathbf{w}^{(l)} := \mathbf{w}^{(l)} - \alpha_0 \cdot \alpha^{(l)} \cdot \mathbf{g}^{(l)},$$

where α_0 is the global LR.

LARS can be used with SGDM or with Adam, known as LAMB [YLR+20]. LAMB was successfully used by Google to train BERT and ResNet-50 with batch size $32K$ with little hyperparameter tuning. The Adam equations are modified as follows in LAMB:

$$\alpha^{(l)} = \frac{||\mathbf{w}^{(l)}||}{||\mathbf{r}^{(l)}||}$$
$$\mathbf{w}^{(l)} := \mathbf{w}^{(l)} - \alpha_0 \cdot \alpha^{(l)} \cdot \mathbf{r}^{(l)}.$$

Other influential optimizers are AdaGrad (in particular, for sparse data), RMSProp, AdaDelta, Nadam, Nesterov accelerated gradient (NAG), AdamW, AMSGrad, and Novo-Grad [DHS11, HSS12, Zei12, Doz16, BLB17, LH19, RKK19, GCH+20]. Figure 4.9 shows an estimated pedigree of optimizers. These are first-order optimizers. AdaHessian is a second-order optimizer that converges to a better minimum than first-order optimizers without the prohibited computational cost of other second-order optimizers [YGS+20]. Given the promising results, AdaHessian adoption may grow.

Stochastic weight averaging (SWA) and LookAhead (LA) are complementary techniques that improve generalization by converging to a better (flatter) minimum [IPG+19, ZLH+19]. The motivation for SWA is that during the later training iterations, SGD bounces between the

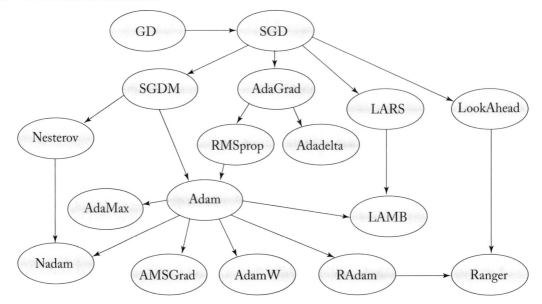

Figure 4.9: A pedigree of optimization algorithms.

borders of a wider minimum. The average of the bounces is a better solution. SWA maintains a separate set of averaged weights $\mathbf{w}_{SWA}$ in addition to the regular set of weights $\mathbf{w}$ used by the optimizer. $\mathbf{w}_{SWA}$ is initialized with $\mathbf{w}$ after completing at least 75% of the training iterations. Then, after completing several iterations, $\mathbf{w}_{SWA}$ is updated as follows:

$$\mathbf{w}_{SWA} := \frac{\mathbf{w}_{SWA} \cdot n_{cycle} + \mathbf{w}}{n_{cycle} + 1},$$

where n_{cycle} is the number of completed cycles after initializing $\mathbf{w}_{SWA}$, and $\mathbf{w}$ is the model learned by the optimizer. One cycle consists of multiple iterations, typically one epoch, but this can vary depending on the dataset's size.

For training, SWA requires $sizeof(\mathbf{w}_{SWA})$ additional memory, which is relatively small compared to the activations and requires negligible additional computations to update. No additional memory or computations is required for serving.

LookAhead (LA) follows a similar approach to SWA [ZLH+19]. The primary difference is that the optimizer updates its weights to $\mathbf{w}_{LA}$ after some iterations: $\mathbf{w} := \mathbf{w}_{LA}$. That is, the moving average $\mathbf{w}_{LA}$ changes the optimization trajectory.

Ranger is a combination of RAdam and LA, and RangerLARS applies LARS techniques to Ranger [Wri19]. We recommend using Ranger as the go-to optimizer and RangerLARS when using large batches.

4.4   BACKPROPAGATION

The rediscovery of the backpropagation algorithm in the 1980s facilitated multilayer NN training. Backpropagation provides an efficient way to compute the gradients, which are then used by the optimization algorithm. This section introduces some of the mathematics behind backpropagation to demystify the learning process; for a reader who may not be interested in all these details, the main takeaway is that backpropagation boils down to multiplications and additions.

The cross-entropy cost function, also known as the log-cost or logistic cost, is as follows:

$$J(\mathbf{w}) = - \sum_{n=0}^{N-1} \sum_{k=0}^{K-1} y_k^{[n]} \log\left(\hat{y}_k^{[n]}\right),$$

where N is the number of samples in a training batch, $y_k^{[n]} \in \{0, 1\}$ is 1 if sample n belongs to class k and 0 otherwise, $\hat{y}_k^{[n]}$ is the model's prediction (as a probability) that sample n belongs to class k. The intuition is that when the model predicts a low probability for the correct class, the cost for that sample is high and vice versa. When $y_k^{[n]} = 1$, as $\hat{y}_k^{[n]}$ approaches zero, the loss approaches infinity. Note that in practice, the cost function includes a weight decay penalty (shown here but often omitted to simplify the notation):

$$J(\mathbf{w}) = - \left(\sum_{n=0}^{N-1} \sum_{k=0}^{K-1} y_k^{[n]} \log\left(\hat{y}_k^{[n]}\right) \right) + \left(\frac{\lambda}{2} \sum_{l=0}^{L-2} \sum_{j=1}^{D^{(l+1)}} \sum_{i=1}^{D^{(l)}} \left(w_{ji}^{(l)}\right)^2 \right),$$

where $\lambda \geq 0$ is the regularization factor.

This objective function is minimized using an optimizer from Section 4.3 chosen by the data-scientist. The input to the optimizer is the gradient or partial derivatives of the cost with respect to each weight $w_{ji}^{(l)}$:

$$\frac{\partial J(\mathbf{w})}{\partial w_{ji}^{(l)}},$$

which needs to be computed for all the weights in a layer and for all the layers of the topology. Each partial derivative is a metric of how a change in the respective weight changes the cost. The optimizer specifies how to nudge each weight to decrease the cost.

Figure 4.10 illustrates how backpropagation works in a toy model to compute one such partial derivative, specifically $\frac{\partial \mathcal{L}}{\partial w_{32}^{(0)}}$, where $\mathcal{L} = J(\mathbf{w})$ to simplify the notation. This partial derivative depends on the next layer's gradient, which depends on the following layer's gradient, and so on. The partial derivative in the color boxes are computed from the forward propagation equations, and their numerical values can be plugged into the chain of equations to determine $\frac{\partial \mathcal{L}}{\partial w_{32}^{(0)}}$. Note that the hidden layer assumes a ReLU activation function. In practice, the partial derivatives for an entire layer are computed as a group using matrix algebra.

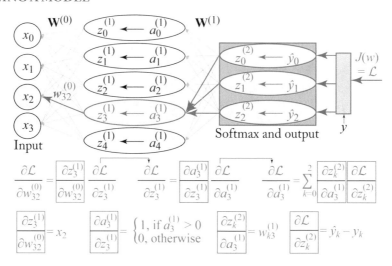

Figure 4.10: Using the chain rule to compute the partial derivative of the cost with respect to a weight in the model. For simplicity, the bias is omitted from the figure.

4.5   TRAINING TECHNIQUES

Training a model involves preparing the dataset and tuning various hyperparameters, such as choosing a topology, selecting an optimizer, and specifying a batch size. In this section, we describe general guidelines in dataset preparation, topology design, and debugging training. These guidelines are based on the current best heuristics rather than a closed-form optimal solution. Thus, experimentation may be required to determine if a guideline is beneficial to a specific training task.

4.5.1   TRAINING DATASET

The first step in training is to manually analyze samples from the dataset to ensure the samples (or most of them) are not corrupted, do not have duplicates, and have proper labels, as well as identify class imbalances. Class imbalances means that training samples are not evenly distributed among the classes. For instance, a dataset used to train a tumor classifier from radiology imagery likely has more images without tumors than with tumors. A simple approach is to oversample the underrepresented class or artificially increase its samples using techniques, such as SMOTE or ADASYN discussed elsewhere [CBH+11, HBG+08], and always analyzing the generated samples to ensure they are realistic. Another approach is to undersample the overrepresented class keeping the harder-to-classify samples. One approach to select the harder-to-classify samples is to train several models, each with a subset of the overrepresented class, and select the misclassified samples. Dealing with class-imbalances is an open research area. Using a metric, such as the $F1$ score, is better than the classification rate for tasks with imbalanced datasets to avoid falling into the accuracy paradox, where the classifier always predicts the over-sampled class. Also, recall from Section 4.2 that class imbalances affect the bias initialization of the last layer.

The training dataset should be split into a training set, validation set (also called out-of-sample or development set), and test set. The training set is used to train the model, and the validation set is used to observe the model's statistical performance on data outside the training set. Hyperparameters are tuned based on the performance on the validation set. The test set should only be used once, after the model and hyperparameters are locked, on data never used to train or tune the model to estimate the performance in deployment. The training-validation-test percentage split depends on the overall size of the training dataset and the similarity between training and deployment data. Assuming all the training samples are from the same distribution, an appropriate percentage split for a dataset with 10,000 samples is 80-15-5, for a dataset with 1 million samples is 95-4-1, and for a dataset with 100 million samples is 99.0-0.9-0.1. The validation and test sets should be sampled from the same distribution as the serving data; that is, they should be as similar to the data used in production to tune the model parameters properly. Any oversampling should be done after splitting the training dataset to avoid data samples being present in both the training and validation sets.

Preprocessing the training set usually involves subtracting the mean and normalizing the variance. It is critical that whatever statistics and techniques used on the training set are also used on the validation set and in deployment. In particular, if the mean of the training set is subtracted from the training samples, then the same *training* mean value should be subtracted in the validation set and in deployment. Mirroring the preprocessing training steps in the deployment data is sometimes ignored when different teams train and deploy models, resulting in lower than expected performance.

Data augmentation is a common technique to increase the dataset size in computer vision, speech recognition, and language processing tasks. In speech recognition, each sample can be augmented by masking or modifying the sample in the time and frequency domain via time and frequency masking and time warping [PCZ+19]. In computer vision, each sample can be left-right flipped, cropped in various locations, and slightly rotated. It is common to augment each sample 10 times, thus, artificially increasing the dataset by an order of magnitude. In language processing, sentences can be duplicated to augment the dataset using the synonyms of some of the words in the sentences.

The last step in preparing the training set is shuffling the order without breaking association with the labels, and manually reanalyzing some of the augmented samples after all the preprocessing steps to ensure they are still valid. Reshuffling the training data between each epoch usually does not help.

4.5.2   DESIGNING A TOPOLOGY

The recommended approach to design a topology for a particular task is to start with a simple topology and then add more complexity. Note that for some tasks, other ML algorithms, such as linear regression or XGBoost (do not worry if you are unfamiliar with these algorithms), which require significantly less compute, may be sufficient.

During the design stage, using *fp*32 and a relatively small batch size ensures that issues encountered are not related to a small numerical representation or a large batch size. Note that as the industry gains more confidence in the robustness of *bf*16, the design stage may shift toward *bf*16. Before increasing the complexity, the designer should verify that the model correctly:

1. consumes the data;

2. generates a valid output;

3. produces the expected cost;

4. learns a better model when trained with real-data vs. random or all-zeros data; and

5. overfits when trained with a tiny dataset, for instance, with two samples.

Then the designer should incrementally increase the complexity with more units and layers, re-verifying the correctness each time. Note that a topology and training process that cannot overfit (the training error is close to or equal to zero) to a few data samples likely indicates a bug.

Section 4.5.3 details the debugging steps when the model is not behaving as expected. The data scientist should monitor the training and validation errors throughout the training process. The training error should decrease by adding more layers and units to the topology; otherwise, this may indicate a bug. This verification-at-every-step approach avoids having a large complex topology full of difficult-to-debug issues. Finding and resolving issues with a smaller topology is significantly easier. Note that introducing a batch normalization layer requires increasing the batch size to ~32 because batch normalization does not work well with tiny batches. A better approach is to use group normalization (see Section 2.6) or another normalization technique which can use micro-batches.

One practical approach is to build up the topology layer-by-layer toward an existing topology (a reference implementation) designed for a related task and dataset size. An alternative approach, detailed in Section 4.5.4, is to start with an existing topology, adapt it to the required task, and tune the hyperparameters. In either approach, the depth and size of the topology depend on the size of the dataset. In both approaches, verification-at-every-step is imperative to a successful design following the debugging steps outlined in Section 4.5.3 when the model is not behaving as expected.

Another recommendation is to incrementally build a deeper model that overfits the training dataset, and then use regularization techniques, such as weight decay, to reduce overfitting. During this process, the data scientist closely monitors the training and validation errors and modifies the topology to decrease the validation error. A high training error indicates the need for a bigger topology. A high validation error indicates the need for regularization or a larger training dataset. Also, the constraints of the serving hardware, such as memory size, should be included in the design process.

Overfitting before regularization serves two purposes. First, it indicates the model is large enough to capture the complexities in the dataset. Second, it is a method to verify the training process is working correctly. Note that data augmentation is a form of regularization reserved for the final design stages.

During the design stage, it is recommended to use the Adam optimizer and a constant LR (as opposed to a decaying LR). More advanced optimizers, such as RangerLARS and advanced LR techniques, such as cyclical LR, should be explored after the topology design is finalized. Note that every step of the design stage may require finding a new LR as deeper models typically need a larger LR.

4.5.3 DEBUGGING TRAINING

Debugging training can be extremely challenging. There are multiple sources of errors in different parts of the training pipeline from the data-processing and topology definition to the optimizer and numerical representation [Kar19]. The following steps can help determine and fix the bug when a model is not training as expected:

1. Use *fp32* to ensure smaller numerical representations are not the cause of the error.

2. Visualize the samples after all the preprocessing steps to ensure no unreasonable distortions were introduced.

3. Verify the validation dataset is preprocessed using the same statistics and techniques as the training set, including the tensor layout.

4. Check that dropout and normalization layers are not simultaneously used; otherwise, permanently remove the dropout layer.

5. Train with a small batch size; if there are batch normalization layers, then use a batch size of ~ 32 or, better, replace batch normalization with group renormalization.

6. Visualize the activation outputs at each layer with a visualization tool, such as Tensorboard, to ensure they make sense; for instance, the first layer in a CNN model typically learns to detect edges.

7. Temporarily reduce the number of training samples to two samples to verify the model can quickly overfit to those training samples.

8. Verify the initial cost matches intuition, for instance, a 0- to 9-digit classification with a balanced dataset should have an initial cost of approximately $-\ln(1/10) \times N = 2.3N$, for a batch of size N.

9. Verify that regular training data results in higher statistical performance than random or zero-input training data; otherwise, this indicates the model is damaging the data or ignoring it.

10. Visualize and look for patterns in mispredicted samples.

11. Use a fixed random seed to exactly reproduce the same behavior when looking for a bug in the code and debug layer-by-layer and op-by-op to find where the observed behavior differs from the expected behavior.

12. Experiment with various weight decay penalties and observe if the training behavior changes as expected: more regularization (a higher penalty) should increase the training error and decrease the test error if the model is overfitting.

13. Experiment with various LRs using both a constant and a cyclical LR, plot the training and validation errors vs. the number of iterations, and observe if the behavior of the errors is as expected.

14. Replace ReLU with LeakyReLU if many gradient values are zero preventing proper learning.

15. Replace all sigmoid functions with hyperbolic tangent functions if the outputs do not have to be between 0 and 1 strictly; limit sigmoid functions to represent probabilities in LSTM gates and for the last layer of a binary classification model.

16. Clip high gradient values.

17. Temporarily remove normalization layers to verify the normalization is not masking some hard-to-find bug.

18. Ensure the correct APIs are used, for instance, the negative log-likelihood loss and the cross-entropy loss are sometimes incorrectly interchanged.

4.5.4 TUNING HYPERPARAMETERS

In this section, we provide recommendations in tuning the hyperparameters: the LR, the batch size, the weight decay, and the optimizer. We also describe how a hyperparameter can affect the other ones. All the recommended hyperparameters require experimentation for proper tuning. Usually, after several training iterations, the set of good hyperparameters narrows to a selected few that can be further narrowed with additional training iterations. That is, full training with every hyperparameter is not required, nor is it practical.

The LR is the most important hyperparameter to tune [Ben12]. There are various techniques for adapting the LR throughout the training process, including the following:

• Constant: uses the same LR for all the iterations.

- Stepwise decreasing: iteratively reduces the LR after a set number of epochs.

- Polynomial decay: slightly reduces the LR in each iteration.

- Cyclical: iteratively decreases and then increases the LR.

- Cyclically decreasing: iteratively decreases the LR for some iterations and resets.

The goal when training a new topology is to achieve a low validation error. A recommended approach to train new topologies is as follows: (1) Use a relatively small batch size (use batch size ~32 if there are batch normalization layers or replace BN with group normalization); (2) Test various initial LRs, such as $10^{\{-5.0,-4.5,\cdots,0.0\}}$, and choose a large enough LR that does not cause training error to diverge [ML18]; (3) Train the model until both training and validation errors flatten [HHS17]; and (4) Decrease the LR by a factor of 10 and return to step (3) repeating several times until decreasing the LR no longer reduces the errors. Optionally, for the last part of the training, switch to a cyclical LR, where the LR decreases and increases again.

The goal when training an established topology with a known validation error is to reduce the training time. The recommendation is to use largest batch size in the batch-size-perfect-scaling region (see Table 4.1). An estimate of this batch size is the sum of the variances for each gradient component divided by the global norm of the gradient. The intuition is that gradients computed with micro-batches have high-variance and vice versa; thus, a good batch size results in the variance of the gradient at the same scale as the gradient itself [MKA+18].

In addition, an initial gradual warmup phase is recommended. If the targeted initial LR is α_0, the optimizer should first use LR of $\alpha_0/20$ and linearly increase this LR over the first ~ 10% epochs until reaching α_0. Then the optimizer should continue with the prescribed LR training technique. The motivation for the warmup phase is to help the training start converging right away with a small LR and then increasing the LR to make faster progress.

For established models, using a polynomial decay LR is a commonly prescribed LR technique:

$$\alpha = \alpha_0 \cdot \left(1 - \frac{t}{T}\right)^2,$$

where α_0 is the initial LR, t is the current iteration, and T is the total number of iterations. Lastly, applying a cyclical LR toward the last ~20% of training epochs can help.

A recommender optimizer is RangerLARS (LARS + RAdam + LookAhead) for large batches and the simpler Ranger (RAdam + LookAhead) for small and medium batch sizes [Wri19].

Another key hyperparameter is the L_2-regularization or weight decay λ. Recommended values to try are $\lambda = 10^{\{-6,-5,-4,-3\}}$. The more a model overfits, the more it requires regularization. Also, other parameters, such as the βs, used in the optimization algorithms in Section 4.3 may require some tuning [Smi17]. Techniques, such as data augmentation, reduced numerical representations (detailed in Section 6.1), weight pruning (detailed in Section 6.3), and larger

LRs contribute to regularization. Using these techniques reduces the required weight decay value. AutoML techniques (introduced in Section 10.1) can also be used for hyperparameter tuning.

4.6   TRANSFER LEARNING VIA FINE-TUNING

Transfer learning via fine-tuning is broadly adopted across many industries. The idea is to use the knowledge gained in a particular *source* task for a different *destination* task. To illustrate, different images have common features starting with edges and growing in complexity. A model can be trained on a large image dataset and then used for another task with a smaller dataset by replacing and *fine-tuning* (retraining) only the upper layers of the model; both tasks can use the same lower level features. The whole model uses the pretrained weights as the initial weights for the nonreplaced layers, and the replaced layers use the traditional weight initialization techniques (discussed in Section 4.2).

Most companies have small datasets compared to the hyperscalers. Fortunately for the community, there are model zoos with models trained with large datasets. Industries and academics with smaller datasets can use these pretrained models and fine-tune them for their related tasks. Fine-tuning existing models dramatically lowers the bar of training large models and drastically increases the adoption of DL.

The following are some guidelines for fine-tuning, and a summary is shown in Figure 4.11.

- Both the source and destination models should share the lower and middle layers; only the upper layers are replaced or reinitialized.

- The number of layers to replace or reinitialize depends on two factors:

  1. the similarities between the source task and the destination task; the more similar the tasks, the fewer layers should be reinitialized; and

  2. the difference between the size of the source and destination dataset; the smaller the difference, the more layers should be replaced or reinitialized.

- Fine-tuning works best when the source dataset is much larger than the destination dataset; if the destination dataset is the same size or bigger, training a new model for the destination task is a better approach.

- The initial LR to fine-tune these models should be 10–100× smaller than the initial LR used to train the original model for the pretrained layers. A regular LR should be used for the replaced or reinitialized layers.

- The same data preprocessing techniques on the original larger dataset should be applied to the datasets used for fine-tuning and validation.

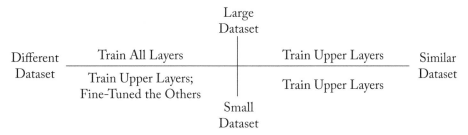

Figure 4.11: High-level guidance on when and what to fine-tune. When the new task's dataset is similar to the original dataset, only the last upper layers should be retrained. When the datasets are different, then training more layers is required. If the new task's dataset is sufficiently large, then it is best to retrain the entire model.

As a simple example, the following steps can be used to design and train a cats vs. dogs classifier (in practice, more recent models have better statistical performance):

1. Replace the last layer of a pretrained VGG16 model from 4096×1000 to 4096×2, as shown in Figure 4.12, since the source dataset has 1000 classes but this task only has 2.

2. Initialize the last layer and use the pretrained weights for the reminder layers.

3. Either freeze or reduce the LR of all the layers except the last one by $100\times$.

4. Train the topology with the target dataset (note that a modern laptop has sufficient computational capacity for this task).

Fine-tuning is also commonly used after making some modifications to the model, such as after pruning or quantizing the weights (discussed in Chapter 6). There are other types of transfer learning techniques, such as domain adaptation, {zero, one, few}-shot learning, and multitask learning [PY10, KL19, WYK+19, Rud17]. These techniques have limited industry adoption.

4.7   TRAINING WITH LIMITED MEMORY

Training requires significantly more memory than serving. During a forward propagation iteration, the activations across all the layers need to be stored to compute the gradients during the backpropagation. Memory capacity can become a bottleneck when training large models, especially on GPUs and accelerators. In this section, we review techniques to mitigate memory bottlenecks.

The most straightforward technique is to reduce the batch size. The size of the activations is proportional to the batch size. However, a batch size less than 32 is not recommended for

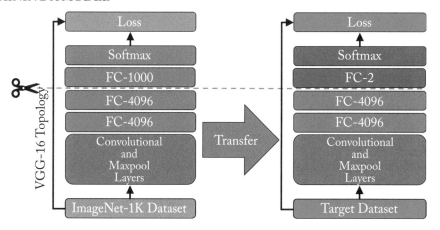

Figure 4.12: Fine-tuning the VGG-16 model for the task of dogs vs. cats classification initially trained on the ImageNet-1K dataset.

models with batch normalization layers. A solution is to replace batch normalization with group normalization technique and use a micro-batch.

The next best technique is *gradient checkpoint* introduced in 2000 and recently gaining traction in academia and some adoption in the industry after the technique resurfaced in 2016 [GW00, CXZ+16]. Gradient checkpoint reduces memory requirements at the expense of additional computations. Rather than storing the activations across all the layers, only the activations of some layers are stored. For instance, a model with 100 layers can have the activations saved every 10 layers. These layers are known as checkpoints, and the group of layers between checkpoints is a segment. During the backpropagation, the activations are recomputed for a particular segment. The process of recomputing them is called *rematerialization*. The activations in memory at a given time are (1) the checkpoint activations and (2) the activations for one segment. In the example with 100 layers and 10 checkpoints, only ~20% of all the activations are stored at any one time. The computation cost is an extra forward propagation. In a GPU or accelerator with high compute capacity and limited memory, this additional compute may require less time and power than storing and fetching the activations from the host.

In practice, uniformly dividing the checkpoints is not a good practice. The total size of the activations and the computational cost of the forward propagation in each segment can significantly vary. Furthermore, checkpoints within skip connections should be avoided. Selecting an optimal number of checkpoint layers that evenly divides the total size of the activations across segments is an NP-complete problem. Jain et al. introduced Checkmate, a system that finds checkpoints for particular hardware targets. Checkmate uses an off-the-shelf mixed-integer linear program solver coupled with a hardware cost model to find suitable checkpoints [JJN+19].

Another technique is to store the activations as 16 bits (as opposed to 32 bits). This reduces the memory and bandwidth usage by up to a factor of 2. NNs are robust to noise, and computing the gradients using activations with half the bits typically does not impact the statistical performance. A related technique is to store compressed activations [JJN+19].

A final technique is deep equilibrium (DEQ), where the depth of the model can vary while keeping the required memory constant. The memory is equivalent to a single layer's activation [BKK19]. DEQ reduces the memory requirements at the expense of additional computations. This technique does not yet have adoption in industry.

In this chapter, we described how to train a model that generalizes and avoids underfitting and overfitting. We explained how to initialize the weights in different layers. We detailed SGD and review various variants. We recommend using Ranger for small to medium batches and RangerLARS for large batches or, for someone new to training, Adam is well documented and simple to get started. We noted that while operating on large batches can result in higher hardware utilization, small batches may generalize better, and we provided guidance on selecting a batch size. We decomposed the backpropagation algorithm as a series of multiplications and additions, which motivate the need for specialized matrix multipliers in hardware. We provided guidelines to topology design and recommended hyperparameters that data scientists should use in the design and debug stage. We explained how to mitigate memory capacity bottlenecks in the training phase at the expense of added compute. For companies with smaller datasets, we recommended modifying an existing model and fine-tuning it for a particular task. In the next chapter, we explore how to accelerate the training by distributing the computations and memory requirements across various compute nodes.

CHAPTER 5

Distributed Training

The number of computations required to train state-of-the-art models is growing exponentially, doubling every ∼3.4 months (far below the glory days of Moore's Law 1.5–2 years) [DH18]. Training a large model can have two primary challenges: (1) the memory required exceeds availability and (2) the time-to-train on a single node can be prohibitively long. To illustrate, training production models commonly used at Google would require 2–16 months on one dedicated DL processor (TPU v2) [JYK+20]. Distributing the computations or the memory requirements among multiple nodes alleviates these challenges and is becoming the norm to train large-scale production models. Hardware designers at Intel, Nvidia, AMD, Google, Graphcore, Cerebras Systems, and others, detailed in Section 7.7, have or are developing dedicated, scalable, multinode training platforms.

Training the popular ResNet-50 model commonly used for image classification requires about 10^{18} (1 exa) operations which is considered small by today's standards and can be trained in under 2 hours with 8 V100 GPUs and in 75 seconds with 2048 V100 GPUs [YZH+18, Nvi20c, YKT+18]. Training the larger 8.3 billion Megatron-LM model requires 12×10^{21} (12 zetta) operations, and can take several days on hundreds of compute nodes [SPP+19]. Training the prodigious 600 billion parameter GShard takes 4 days on 2048 TPU v3 accelerators [LLX+20].

The main techniques to distribute a training workload across multiple nodes are data parallelism and model parallelism (including pipeline parallelism), illustrated in Figure 5.1, and a hybrid of these. Also, federated learning is a form of data parallelism distributed training in edge (client/IoT) devices. Data and model parallelism benefit from high bandwidth interconnects between the nodes. In data parallelism, a batch (called the global-batch in this chapter) is split among the worker nodes and called the node-batch, with each node working on the same model. The nodes communicate the weight updates. In model parallelism, the model is split among the worker nodes, and the nodes communicate the activations. Model parallelism is typically used when the memory requirement exceeds the node's memory. In hybrid parallelism, data parallelism is used across groups of nodes (super-nodes), and model parallelism is used within each super-node.

Data parallelism is more commonly used in industry, but as the sizes of the models are growing, hybrid parallelism is becoming the norm for state-of-the-art models. In the remainder of this chapter, we describe data and model parallelism, their typical usages in data center training, and their limitations. We also discuss federated learning, and we review various communication primitives.

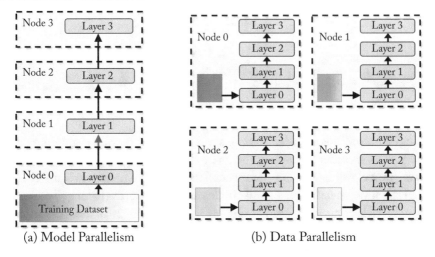

(a) Model Parallelism   (b) Data Parallelism

Figure 5.1: (a) In model parallelism the model is distributed among multiple compute nodes. (b) In data parallelism, the training dataset is split among multiple compute nodes and each node has the entire model.

5.1 DATA PARALLELISM

Distributed training using data parallelism is widely adopted at hyperscalers to reduce the total time-to-train (TTT). It is the basis for *federated learning*, detailed in Section 5.3, used for edge device distributed training. In data parallelism, each batch is split among the worker nodes. Each worker node computes the gradient with respect to its node-batch and communicates it to a root node or, when no root node is used, with the other worker nodes.

Synchronous SGD (Sync SGD) requires that all the worker nodes have the same weights at the end of each iteration. Using Sync SGD across various nodes results in precisely the same trained model as using SGD on one node with the global-batch, that is, with a batch that combines all the node-batches. A simple communication strategy is for each worker node to communicate the node gradient to a root node known as the *parameter server* (PS). The PS aggregates the node gradients, updates the global model weights, and broadcasts (sends) the updated global model to all the worker nodes. The main drawback is the synchronization overhead from waiting to receive the node gradients, in particular, due to straggling (slowest) workers or slower network links. In practice, more advanced communication strategies are used and discussed in Section 5.4.

Asynchronous SGD (Async SGD) does not enforce synchronization before updating the global model and alleviates the synchronization overhead in Sync SGD. However, Async SGD has other challenges. Async SGD requires more tuning of the optimization hyperparameters (such as the momentum constant), requires more iterations to train, and typically has worse convergence performance than Sync SGD when the number of worker nodes is higher than 8 [MZH+16]. The convergence is affected because the global model receives stalled (older and less relevant) node gradients. Furthermore, Async SGD does not match single-node SGD performance, and therefore it is more difficult to debug. Async SGD (unlike Sync SGD) typically does not scale past 8 (and often 2 or 4 is the limit) nodes. An Async–Sync SGD hybrid approach can work where the nodes are clustered in up to 8 groups. Updates within a group are synchronous, and updates between groups are asynchronous. Going beyond 8 groups reduces performance. The main exception where Async is used across several nodes with minimal impact is for the embedding layer in large recommender systems (and sometimes the layer immediately after) as the updates only affect a limited subset of embedding entries. Note that in the Sync SGD and Async SGD literature and in this section, the term *SGD* is typically an alias for all the extensions of SGD discussed in Section 4.3, such as SGDM, Adam, and LARS. The rest of this section focuses on Sync SGD.

Ideally, the TTT is reduced proportionally with the increased number of compute nodes known as perfect linear scaling. That is, doubling the number of nodes halves the TTT. However, there are algorithmic and communication challenges that prevent perfect scaling:

1. Large batches require a few additional training epochs to converge to a flat minimum.

2. There is a limit on the maximum batch size that decreases the training iterations.

3. There is higher communication overhead with more nodes.

The methodologies below mitigate some of these challenges.

A large global-batch is essential so that the node-batch is sufficiently large for high compute utilization among the worker nodes. However, there is a limit on the maximum batch size that decreases the training iterations. Experimentation with careful LR turning is required to find that batch size, as discussed in Sections 4.3 and 4.5.4. Also, using an optimizer more robust to large-batches, such as LAMB, and a warmup phase is recommended.

Communication overhead also hurts the scaling performance: the more nodes and more data, the higher the communication time. After a point, the execution time is dominated by communication time and cancels out the savings from the additional compute. The following reduces this overhead (these techniques also apply to federated learning):

1. hiding (overlapping) the communication between nodes by the computation within a node (node-to-node high-bandwidth is beneficial);

2. compressing the gradients before communicating them (related work used 1 bit and 2 bits to communicate the gradients) [LHM+18, SFD+14, WXY+17];

3. computing more local iterations before synchronizing;

4. ignoring stagnant nodes and updating the global gradient using ∼95% of the nodes (this is not possible with some communication primitives); and

5. sparsifying the gradients, that is, only transmitting the largest magnitude gradients and accumulating the rest locality [AHJ+18].

You et al. achieved extensive scaling using some of these techniques. They partitioned a $32K$ batch size using $1K$ CPU nodes and achieved the fastest ResNet-50 TTT at the time [YZH+18]. Similarly, You et al. achieved record scaling and TTT on the BERT model on a TPUv3 Pod (1024 chips) [YLR+20]. Today, ResNet-50 can be trained in several days on one V100 GPU node or in ~2 minutes (~1 epoch per second on ImageNet-1k) using 3456 V100 GPU nodes or using a TPUv3 Pod with no accuracy drop [MSU+19, YKC+18].

5.2 MODEL PARALLELISM

The adoption of model parallelism in production is expanding as the size of the models (and the size of the embedding layers in recommenders) continues to grow, and the memory required for training transcends the capacity of a single GPU or accelerator node. Today, models with hundreds of millions of weights are common in production, and new models are approaching one trillion weights [LLX+20].

In model parallelism, the model (the weights) is partitioned and distributed among the nodes. There are two main partitioned strategies: parallel layers and sequential (pipeline) layers, the latter one shown in Figure 5.1b. In both strategies, each node processes the entire batch and communicates the activation in the forward propagation and the activation gradients in the backward propagation. Partitioning the layers with a balanced load (known as *device placement*) is an NP-complete problem. Dynamic programming and RL can improve device placement [HNP+18, MGP+18]. Optimal device placement is a topic of ongoing research. Other less common strategies are partition across channels or partition across spatial axes in CNNs using a modified convolution operator [GAJ+18, DMM+19, JZA18, JGK18].

Model parallelism with parallel layers runs simultaneously on separate nodes and occasionally communicate their activations. The original design of the legendary AlexNet topology used this approach with two GPUs to mitigate the memory limitations of a single GPU [KSH12]. The current model parallelism algorithms can often scale up to 4 or 8 nodes with careful tuning. Embedding layers in recommender systems, such as Facebook's DLRMs, can scale to more nodes [NKM+20].

When dozens or hundreds of nodes are available to train large size models, a hybrid model–data parallelism can be optimal. This hybrid approach was used by Google to train the 11 billion weights T5-Transformer on TPUs-v3 and by Nvidia to train the 8.3 billion weights Megatron-LM model on V100 GPUs [RSR+19, SPP+19]. The Mesh TensorFlow (mtf) library supports hybrid parallelism and uses an integer linear programming (ILP) algorithm to determine the partition [SCP+18].

5.2.1   PIPELINE PARALLELISM

Pipeline parallelism is a type of model parallelism (although some literature treats it as separate). Each layer or group of sequential layers is placed on a separate node. Consecutive batches are put into the pipeline to keep it full. The communication between the nodes is limited to the activations of the layers where the partitions occur and their respective activation gradients. The downside of pipeline parallelism is that the updates use stalled weights with similar issues as Async SGD. Chen et al. proposed using the momentum to estimate the weight and scaled to 4 GPU nodes with no convergence degradation [CYC19]. Cerebras System proposed a linear predictor to mitigate staleness and scaled to 169 nodes on the ResNet-110 model with the CIFAR-10 dataset, a small research dataset [KCV+20]. Additional algorithmic advancements are required to mature pipeline parallelism to scale to hundreds of nodes on production models and datasets.

There is limited support for efficient pipeline parallelism (and model parallelism in general) in the major frameworks. To improve pipelining, Google introduced GPipe [HCB+19]. GPipe splits a batch into micro-batches to reduce idle time in the model pipeline and accumulates the gradients across the micro-batches to maintain statistical consistency. The user specifies how to partition the model, that is, which portions of the model are allocated to which nodes. Similarly, Microsoft uses PipeDream and DeepSpeed for GPU pipelining [HNP+18]. Graphcore supports pipelining with gradient checkpoint (discussed in Section 4.7).

5.3   FEDERATED LEARNING

Federated learning is a decentralized learning methodology introduced by Google and is an extension to data parallelism [MMR+17]. The goal of federated learning is to use a large number of local client devices, such as mobile phones, to train a model without transmitting the training data. Federated learning is gaining adoption in the industry due to the data privacy and network bandwidth benefits. It is used to train the word predictor in Google Gboard and URL predictor in Firefox [HRM+19, Har18].

In federated learning, as illustrated in Figure 5.2, an untrained or partially trained global model is pushed from a centralized global server to a large number of local client devices. Each device trains the model for multiple epochs using its local data and then transmits the updated local model to the central server within a given time window. All the devices train for the same local epochs and with the same batch size. The number of local iterations per epoch varies as it depends on the size of the local training dataset. The global server updates the global model as an average of the local models (more on this below). The server then broadcasts this new global model back to the client devices, and the process repeats. The number of epochs, batch size, and number of clients requires tuning for best performance.

Federated learning is especially useful when the client device has sensitive or private data the user does not want to share with an external server, or when the cost or the power to transmit the training data is higher than processing locally. Examples include mobile phone apps, health-

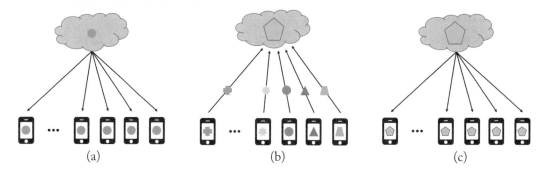

Figure 5.2: Federated learning. (a) An untrained global model (represented by the green dot) is broadcasted to the client devices. (b) The training happens in each client device, and only the updated client model (represented by the various geometric figures) is transmitted to the cloud to update the global model. (c) The updated global model (represented by the gray pentagon) is broadcasted to all the client devices, and the process repeats. Based on [Sas19].

care patient data, companies' emails, and manufacturing equipment data. Some organizations (e.g., a hospital) can be thought of as a device client among a group (e.g., a group of hospitals) in federated learning.

Federated learning is a generalized form of Sync SGD but, rather than synchronizing after every iteration, the weights are synchronized after some local epochs. The more infrequent the synchronizations, the more likely the model has convergence challenges. However, frequent synchronizations consume significant network bandwidth, which is prohibited in some devices. The primary challenge of federated learning is to reduce the synchronization frequency (by increasing the number of local epochs) and maintain the expected training convergence.

Two additional challenges can affect convergence. First, the data in each device is typically not independent and identically distributed (IID); data within a client is more similar than data across clients, and the number of samples between clients varies. This non-IID violates the guidance to randomize the order of the samples in the training dataset so each batch has IID samples.

Second, the local devices have heterogeneity both in computational capacity and network reliability across devices. In particular, mobile phones vary significantly in memory, compute, and network connectivity with approximately two-thirds of operating mobile phones in the world being over six years old [Haz18].

The server uses an average of the local models weighted by the number of training samples in each device to compute a global model update. Alternatively, a more stable approach is to randomly choose the clients (assuming a large pool of candidates) with probability proportional to the number of training samples in each device, and use an unweighted average to compute the global model update [LSZ+19].

A federated learning system uses more clients than needed to train local models to mitigate device and network unreliability. A system may assign 300 devices to train local models but only needs to collect local models from 256 devices. Assuming each device uses a local batch size of 16, then the global batch size is $256 \times 16 = 4096$, which may be the limit (the largest batch size that converges to an adequate minimum) for some topologies.

A simple technique to improve robustness to both non-IID batches and local models that are unable to complete the local number of epochs is to use a proximal term. This term is a small adaptable penalty in the objective function for significant deviations from the global model. Note that it is better to communicate a local model that has not completed the requested epochs than to ignore it [LSZ+19].

Communication overhead can be reduced by quantizing with rotations and communicating the weight changes [KMY+17]. A randomly applied mask can further reduce the number of communicated parameters. Traditional data compression techniques can also be used. These techniques also apply to conventional Sync SGD data parallelism to decrease network traffic but are more critical in federated learning due to the higher communication cost. Optimization techniques, such as LAMB and RangerLARS, used in data centers, can be applied to federated learning to increase the number of client devices and accelerate training. Also, TensorFlow provides an API to simulate federated learning with a couple of additional lines of code.

Areas of Caution
Three areas of caution are as follows:

1. Training and communicating a model can be expensive (in terms of battery and data consumption). These expenses are mitigated by limiting training to periods when the device is plugged in and idled and communicating the local model when the device is on a free wireless connection.

2. Despite not transmitting the training data, some information about the local training data can be extracted from local models [HAP17]. To preserve privacy, for instance, Google uses secure aggregation where the local models are only unencrypted and averaged when multiple models become available to the server [BIK+17]. OpenMined developed PySyft on top of PyTorch to improve privacy. Section 10.3 discusses other ongoing work to maintain privacy.

3. Older devices with limited computational and memory capacities, and devices in remote areas may not proportionally contribute to the overall training. This imbalance results in a model that learns characteristics biased toward more affluent populations. Further work is required to mitigate this.

5.4   COLLECTIVE COMMUNICATION PRIMITIVES

There are various communication functions, known as collective communication primitives, and library implementations. These primitives are used in data parallelism to communicate and then aggregate the local gradients, in model parallelism to communicate the activations and their respective gradients, and in transitioning between model and data parallelism to rearrange the data properly. Some common collective communication primitives are as follows:

- **Broadcast**: M elements in the root node are copied to the other $P-1$ processor nodes, as shown in Figure 5.3a.

- **Scatter**: M elements in the root node are partitioned, and each partition with $M/(P-1)$ elements is copied to a different processor node, as shown in Figure 5.3b.

- **Reduce**: the root node receives M elements from each of the others $P-1$ processor nodes and performs a reduction operation, such as sum, maximum, minimum, mean, or product, across each of the $P-1$ elements.

- **Gather**: the root node receives $M/(P-1)$ elements from each of the other $P-1$ processor nodes and concatenates them (equivalent to Figure 5.3b with the arrows reversed).

- **AllReduce**: Equivalent result as Reduce + Broadcast.

- **AllToAll**: M elements in each node are partitioned, and each partition with $M/(P-1)$ elements is copied to a different processor node where the received partitions are concatenated. Equivalent result to Scatter and Gather for all nodes, as shown in Figure 5.3c.

- **AllGather**: Equivalent result as Gather + Broadcast.

- **ReduceScatter**: Equivalent result as Reduce + Scatter.

The AllReduce, AllToAll, and AllGather primitives do not require a dedicated root node. While their end-result is equivalent to sequentially using two simpler primitives, they typically use more efficient implementations. Later in this section, we analyze various AllReduce implementations.

The MPICH, OpenMPI, Intel MPI, and MVAPICH libraries implement primitives using the Message Passing Interface (MPI) standard specifications. The MPI is a library specification that operates at the transport layer implemented by MPICH and other libraries in C/C++ and Fortran with message-passing standards and APIs. In the MPI specification, each processor node has a unique address space. The literature on collective communication primitives is extensive, including their optimizations for clusters connected by switched networks and a study of MPI usages [TRG05, LMM+19].

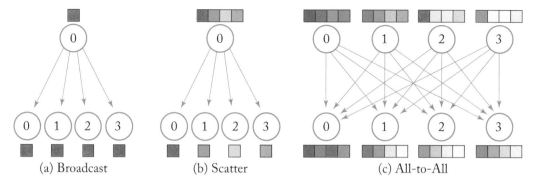

Figure 5.3: (a) The broadcast primitive copies a set of elements in the root node to the other nodes. (b) The scatter primitive copies a separate partition of a set of elements in the root node to the other nodes. Note that reversing the arrows results in the gather primitive. (c) The all-to-all primitive (also known as transpose) copies a separate partition of a set of elements in each node to the other nodes, where the received partitions are concatenated.

Libraries that offer higher-level communication functions using existing primitives libraries or reimplementing them are: Horovod, Nvidia's NCCL, Facebook's Gloo, Intel's oneCCL, and SparCML and Blink from academia [SDB18, RAA+19, WVP+19]. Horovod has broad industry adoption for GPU and CPU distributed training. It is supported by various DL libraries, including TensorFlow, PyTorch, and MXNet. Horovod uses NCCL for GPUs and oneCCL, MPI, and Gloo for CPUs. Uber developed and contributed Horovod to the LF AI foundation.

The most common primitives used in distributed training are (in this order) AllReduce, AllToAll, and AllGather. AllReduce is used to aggregate the local gradients in data parallelism. AllToAll is used to exchange the activations and activation gradients in model parallelism and to transition from model to data parallelism. AllGather is used to concatenate activations or gradients in a specified order, for instance, in Gshard to change a sharded (broken) tensor to a replicated tensor [LLX+20].

In Sync SGD data parallelism, the end result of AllReduce is for all the nodes to receive the aggregated sum of all the local weight gradients; that is, the reduction happens across the nodes. For instance, during the backpropagation of a typical convolution layer with a 4D weight gradient tensor (number of kernels, number of channels, kernel height, and kernel width), the AllReduce primitive aggregates the 4D tensors across all the nodes and broadcasts the sum. In Sync SGD, AllReduce is necessary to ensure the weights across all the nodes are the same at the end of each training iteration. AllReduce algorithms differ in the specific mechanism to achieve this Reduce+Broadcast, but the results are the same.

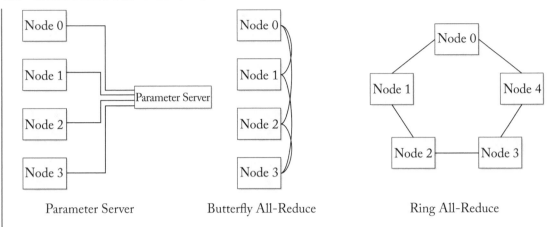

Figure 5.4: Various AllReduce algorithms.

In the following analysis, we examine four AllReduce algorithms based on the number of nodes, latency, and bandwidth: parameter server (PS), AllReduce-Ring, AllReduce-Butterfly, and AllReduce-Tree, shown in Figure 5.4. We assume there are P nodes connected in a 1-hop all-to-all (fully connected) physical network: each node-to-node link has the same latency L independent of how many nodes are communicating. We also assume the links are bidirectional with a per directional bandwidth of B between any two nodes, and the nodes can simultaneously send and receive messages without affecting the unidirectional performance. The terms *node* and *processor* are used interchangeably, and *rank* refers to the node ID from 0 to $P - 1$. Note that the physical network topology impacts which algorithm is optimal. For instance, running an AllReduce-Ring algorithm on a system with a ring physical network topology is much better than an AllReduce-Butterfly algorithm on the same ring physical topology since the load would not be balanced between links. Section 7.5 discusses the physical interconnects and physical network topologies.

Note the difference between network latency and bandwidth. The latency L is the time to communicate one byte from one node to another. The bandwidth B is the number of bytes that can move through the network per second (the width of the network pipeline) per direction. The total execution time T to transfer a message of M bytes from one node to another node is:

$$T = L + M/B = L + T',$$

where $T' = M/B$ is the time it takes to move the data without accounting for the latency. The above equation ignores the software overhead and the time to aggregate (sum) the M elements by the receiver node.

PS performs a reduce-sum and then a broadcast operation, which requires two steps. The total execution time is:

$$T_{PS} = T_{reduce} + T_{bcast}$$
$$= (L + T') + (L + T')$$
$$= 2 \cdot (L + T').$$

The data moves in one direction, and most of the links in the fully connected physical network are unused.

The AllReduce-Ring requires two steps in a 1-hop all-to-all physical network. In step 1, each node breaks down the message into P smaller packages and sends a message of size M/P to each of the other $P - 1$ nodes, and the receiver nodes aggregates the messages. In step 2, each node broadcasts the aggregated message of size M/P to each of the other $P - 1$ nodes. The total execution time is:

$$T_{ring} = 2 \cdot (L + T'/P).$$

The data moves bidirectionally using all the links in the fully connected physical network.

The AllReduce-Tree performs a reduction and a broadcast operation both in a tree pattern, which requires $2 \cdot \log(P)$ steps (log is base 2 and a floor operator). The total execution time is:

$$T_{tree} = 2 \cdot \log(P) \cdot (L + T').$$

Using two trees simultaneously in each link reduces the time, with each tree working on half the data. Each package is of size $M/2$. A similar approach is the Two-Tree algorithm, also known as Double Binary Tree [SST09]. The total execution time using bidirectional links is:

$$T_{tree2} = 2 \cdot \log(P) \cdot (L + T'/2).$$

Most of the links in the fully connected, physical network are unused.

The AllReduce-Butterfly requires $\log(P)$ steps. For simplicity, we assume P is a power of 2. During each step, a package is exchanged with a neighbor in a butterfly pattern. More precisely, at step $s \in [0, \log(P) - 1]$, node $p \in [0, P - 1]$ sends and receives a package of size M to node $p + \frac{P}{2^{s+1}} < P$. The total execution time using bidirectional links is:

$$T_{bf} = \log(P) \cdot (L + T').$$

The analysis shows that for homogeneous all-to-all physical topology, the AllReduce-Ring has the lowest execution time when $P > 2$. This homogeneity is typical for 1-hop connections, where two nodes only go through one network switch to communicate, such as a rack of CPUs or a DGX-2 system. Most CPU rack designs rely on the top-of-rack (ToR) switch even for intra-chassis CPU message passing. For chassis with internal switches, the analysis above only applies to CPUs within the chassis. In a DGX-2 system, GPU nodes have 300 GB/s bidirectional

NVLink links (150 GB/s in each direction) (note that the GPU nodes in a DGX-2 system have an additional, albeit smaller, 32 GB/s bidirectional link through PCIe).

Large-scale distributed training across nodes that require multiple hops usually involves multiple communication primitives. Otherwise, the largest latency and smallest bandwidth link would determine the primitive's latency and bandwidth. A common approach to scale, for instance, across multiple DGX-2 systems is to use AllReduce-Ring within each DGX-2, then AllReduce-Ring across the DGX-2 systems, and then broadcast within each DGX-2. A similar approach can be employed with racks of CPU servers.

Wang et al. developed a collective communication library known as Bling that efficiently uses heterogeneous links [WVP+19]. Bling uses a collection of spanning trees to find various paths to pass messages in parallel and has shown to outperform other libraries in the presence of heterogeneous network links.

In this chapter, we addressed three challenges to training some models: the required memory exceeds availability, the time-to-train is prohibitively long, and the training data is scattered across multiple edge devices. We detailed data and model parallelism. Data parallelism is more commonly used in industry and is supported by the major frameworks. However, some impediments include memory constraints for prodigious models, high communication latency for large models, large global-batch to scale, and small node-batch inefficiencies. Model parallelism can be used for large models, but usually, the scaling is limited to eight nodes, the optimal way to split the model is an NP-complete problem. There is limited support in the major frameworks for efficient model parallelism. Pipeline parallelism suffers from stalled weights, and we discussed some work to partially mitigate this. Hybrid parallelism is becoming the norm for state-of-the-art models. Data parallelism is used across groups of super-nodes, and model parallelism is used within each super-node with 4–8 nodes per super-node. In the next chapter, we explore the various formats to represent numerical values used in production and those in academic exploration as well as compression techniques to reduce the memory footprint of models.

CHAPTER 6

Reducing the Model Size

Computers represent real numerical values as a set of binary digits or bits, usually with 8, 16, 32, or 64 bits. The more bits used, the higher the numerical range and precision or representation of the numerical value. The numerical format of a model can impact its computational and statistical performance. Using a smaller numerical representation can increase the number of operations per cycle and reduce memory, memory bandwidth, network bandwidth, and power consumption. In particular, if a workload is memory bandwidth bound (bottlenecked by the memory bandwidth), reducing the numerical representation alleviates such bottleneck and improves the computational performance. If it is compute bound (bottlenecked by the compute available), hardware designers can pack more smaller numerical format multipliers into a given die area to improve the computational performance. However, using a smaller numerical representation may result in lower statistical performance for some models.

Figure 1.17 shows various numerical formats with the respective number of sign, exponent, and mantissa bits. The exponent bits determine the range, and the mantissa bits determine the precision. For instance, *fp*32 and *bf*16 have the same range factor, but *fp*32 provides higher precision.

There are four main techniques used to reduce the model size:

1. reducing the numerical representation;

2. pruning (trimming) parts of the model and compressing the pruned model;

3. distilling the knowledge to a smaller model; and

4. using NAS that rewards small models.

While most commercial applications use *fp*32 for training and inference workloads, lower numerical formats are rapidly gaining adoption. Specifically half-precision floating-point (*fp*16) and bfloat16 (*bf*16) for training and inference and, for a subset of workloads, *int*8 for inference, all with 32 bits accumulation for MAC operations. Using *bf*16 or *fp*16 multipliers with *fp*32 accumulators has insignificant to no loss in the accuracy for training and inference. Using *int*8 multipliers with *int*32 accumulators has some to minimal loss in the accuracy for some inference workloads. Note that storing the activations in a 16-bit format reduces memory and bandwidth consumption by almost 2×, even if the hardware does not support 16-bit multiplies [Dev17].

Training requires a larger numerical representation than inference, in particular, to capture the dynamic range of the gradients and weight updates. Figure 6.1 shows the histogram of

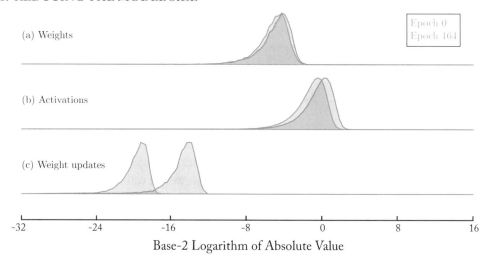

Figure 6.1: Distributions of the ResNet-110 weights, activations, and weight updates at two separate training epochs using the CIFAR dataset. Adapted from [KWW+17] with the authors' permission.

log-base 2 absolute values from ResNet-110 tensors across two separate training epochs and illustrates the larger range of the weight update values.

An active research area is to develop numerical representations that better represent the values with 8 bits and 4 bits and are simple to implement in silicon. Using a smaller numerical representation can improve training and inference even if the hardware does not support higher peak operations per cycle at the smaller representation because the memory bandwidth savings accelerate memory bandwidth bound layers, which are common.

Models are typically overparameterized, which facilitates training and provides opportunities to reduce the model size post-training. Trained models typically have several small weights. Forcing them to zero can have computational advantages with minimal to no statistical impact. This process is called *pruning* and results in a sparse model. There are two types of model sparsity, discussed in Section 6.3, structured and unstructured.

A key benefit of sparse models is improved compression. Compression reduces the memory footprint and memory bandwidth consumption at the expense of some additional computations for decompression. The time for this additional decompression is usually less than the additional time to transmit the uncompressed data; therefore, compression is advantageous.

A small model can be trained to produce the output of a large trained model. The knowledge of the larger trained model (the teacher model) is distilled to the smaller model (the student model). This method is known as knowledge distillation.

In Section 6.1, we review the various 16-bit and 8-bit numerical formats adopted in production, as well as other promising formats. In Section 6.2, we discuss techniques to quantize a model from *fp*32 to *int*8. In Section 6.3, we review pruning and compression techniques. In Section 6.4, we explain knowledge distillation in more detail.

6.1   NUMERICAL FORMATS

The most popular and widely adopted format is *fp*32 for both training and inference. The industry is moving toward *fp*16 and *bf*16 for training and inference, and for a subset of workloads, *int*8 for inference. Nvidia introduced a nonstandard *fp*19 format (sometimes referred to as bfloat19) for matrix multiplications, which combines the range of *bf*16 and the precision of *fp*16. Intel and IBM explored nonstandard *fp*8 formats. Figure 1.17 shows various numerical formats with the respective number of sign, exponent, and mantissa bits. The mantissa is also known as the *significand* and should not be confused with the term *mantissa* used in the logarithmic literature to refer to the fractional part of a logarithm.

Looking ahead for different hardware usages, the numerical formats that are or can be used across various types of development stages are:

- topology research and topology design: *fp*32;

- training production models in data centers: *fp*32, *bf*16, *fp*16, and *fp*19; limited *fp*8;

- serving production models in data centers: *fp*16, *bf*16, and *fp*8; some *int*8; extremely limited *int*4; and

- serving production models in edge devices: *fp*16 (depending on power constraints), *int*8, and *fp*8; some *int*4.

DL libraries, such as TensorFlow, PyTorch, MXNet, OpenVINO, and TensorRT, support *int*8, *fp*16, *bf*16, and *fp*32. For other formats to gain adoption, hardware and framework support is needed.

Table 6.1 shows the range, the minimum and maximum positive values for the floating-point numbers, and the maximum numerical error across various numerical formats. *fp8-ibm* refers to an 8-bit floating-point format introduced by IBM and discussed below. $u\{4, 8\}$ represents a $\{4, 8\}$-bit unsigned integer, $s\{4, 8, 16, 32\}$ represents a $\{4, 8, 16, 32\}$-bit signed integer, and (n_S, n_E, n_M) indicates the number of sign, exponent, and mantissa bits, respectively, of the floating-point formats. Thus, $(1, 8, 23)$ indicates a format with a sign bit, 8 exponent bits, and 23 mantissa bits, which corresponds to *fp*32. The exponent bits determine the range and the mantissa bits the precision. The maximum numerical error of a given floating-point representation is the floating-point number multiplied by

$$1/2^{(n_M+1)}$$

Table 6.1: A comparison of different numerical formats. The maximum numerical error of a given floating-point representation is the floating-point number multiplied by *Maximun Error*.

Format (n_S, n_E, n_M)	Positive Range	Positive Minimum	Positive Maximum	Maximum Error
*fp*32 (1, 8, 23)	$[2^{-126}, 2^{128}]$	1.4×10^{-45}	3.40×10^{38}	6.0×10^{-8}
*fp*19 (1, 8, 10)	$[2^{-126}, 2^{128}]$	1.1×10^{-41}	3.40×10^{38}	4.9×10^{-4}
*bf*16 (1, 8, 7)	$[2^{-126}, 2^{128}]$	9.1×10^{-41}	3.39×10^{38}	3.9×10^{-3}
*fp*16 (1, 5, 10)	$[2^{-14}, 2^{16}]$	6.0×10^{-8}	65504	4.9×10^{-4}
*fp*8 (1, 5, 2)	$[2^{-14}, 2^{16}]$	1.5×10^{-5}	57344	0.125
*fp*8 (1, 4, 3)	$[2^{-6}, 2^{8}]$	2.0×10^{-3}	240	0.0625
*fp*8–*ibm* (1, 4, 3)	$[2^{-10}, 2^{4}]$	1.2×10^{-4}	15	0.0625
*s*32	$[1, 2^{31} - 1]$	1	2.15×10^{9}	0.5
*s*16	$[1, 2^{15} - 1]$	1	32767	0.5
*s*8	$[1, 2^{7} - 1]$	1	127	0.5
*u*8	$[1, 2^{8} - 1]$	1	255	0.5
*s*4	$[1, 2^{3} - 1]$	1	7	0.5
*u*4	$[1, 2^{4} - 1]$	1	15	0.5

or 0.5 for the integer representations.

Training a model with 16 bits (specifically *bf*16 or *fp*16) usually requires the following:

- MAC operators with 16-bit operands accumulated to *fp*32, and the accumulation is converted to 16-bit after totalling the running sum (note that the hardware logic may accumulate to less-bits registers, such as $(1, 8, 21)$ to reduce cost);

- reductions (sums) accumulated to *fp*32 and the result converted to 16-bit;

- activation functions at either *fp*32 or 16-bit;

- activations stored in 16-bit;

- a copy of *fp*32 weights used for the weight update (the updates use 16-bit gradients); and

- a copy of the updated weights converted to 16-bit for the next iteration.

The first three bullets also apply to inference with a 16-bit or 8-bit format. In both cases, accumulation to a larger numerical format is recommended to avoid numerical overflow (notation: MAC source → MAC destination): {*fp*16, *bf*16} → *fp*32, and *int*8 → *s*32 (signed *int*32).

Floating-point 16-bit bfloat (*bf*16) was introduced by Google as brain floating-point. Models are robust to additive noise, and, in fact, it is a common practice to add noise when training a model in the form of weight decay regularization, as discussed in Section 4.1. Reducing the mantissa bits from 23 in *fp*32 to 7 in *bf*16 can be interpreted as injecting noise into the model. *bf*16 maintains the same range factor as *fp*32 and is particularly useful to support the range in the gradients. Experiments demonstrate that models trained with *bf*16 have virtually the same accuracy as those trained with *fp*32 with the same number of iterations, without changing any hyperparameter, and without scaling the objective function cost [KMM+19]. However, there may be outlier models where these observations are not valid. Also, when the number of classes is greater than 2^{n_M} or 127, *fp*32 should be used for the cost function. Moreover, while softmax alone can use *bf*16, various implementations combine the softmax function and the cost function. Those implementations should use *fp*32.

While *bf*16 was primarily designed for training (the large exponent to represent the gradients), it is also used for inference with similar computational gains over *fp*32. Google TPU v2–4, the Habana Gaudi AI processor, the 3rd-generation Intel Xeon Scalable processor (codename Cooper Lake), the Arm-based Neoverse N2 "Zeus" CPU, and the Nvidia A100 GPU have *bf*16 multipliers.

Floating-point 16-bit half-precision (*fp*16) is used for inference and training, the latter often requiring a technique known as *loss-scaling*. During training, particularly during the early stages, the magnitude of many activation gradients often falls below the supported range of *fp*16 and gets truncated to zero and the upper range of *fp*16 is unutilized. Scaling the loss (more precisely, the cost or objective function), mitigates this inability to represent very small values and enables the use of the higher range. Specifically, the cost is scaled by a value $\gg 1$ without overflowing the activation gradients past the upper *fp*16 range. Then, unscaling the weight gradients by the same factor before the weight update. In addition, normalizing 0–255 RGB input image value to 0–1 and adding batch normalization to the activation reduces overflow risks [Wu19]. Nvidia GPUs, AMD Radeon GPUs, Huawei Atlas and Ascend processors, and Graphcore Colossus have *fp*16 multipliers.

The primary advantage of *bf*16 over *fp*16 is avoiding the need to implement loss-scaling, which requires empirical tuning. This advantage is particularly significant for models requiring dynamic loss scaling (and dynamic tuning) such as GNMT and Transformer, given the large variations in gradient distribution throughout training, which increases the software complexity [MSD+19]. Some tools, such as OpenSeq2Seq, can automate dynamic loss scaling for some models [KGG+18].

A disadvantage of *bf*16 over *fp*16 is the 3 fewer mantissa bits; there may be some precision-sensitive workloads that benefit from those bits. The upper range values of *bf*16 are not used, bringing to question the need for 8 exponent bits for most training workloads. Facebook, for instance, uses *fp*16 (rather than *bf*16) to store the embedding layers (not for MAC operators) in DLRM training (the MAC operators of the embedding layers happen in *fp*32) [ZYY18]. In designing a training processor, it is recommended to support both *fp*16 and *bf*16 (using a 19-bit (1, 8, 10) *fp*19 floating-point circuitry unit) to facilitate transitioning from existing hardware that only support one format (*fp*16 or *bf*16).

TensorFloat-32 with 19-bit floats (*tf*32) was introduced by Nvidia starting in the Ampere architecture. TensorFloat-32 uses *fp*19 MACs with *fp*32 accumulation. All the operations and storage happen in *fp*32 except for the MAC operations used in matrix multiplications. Those *fp*32 MACs are replaced with *fp*19 MACs and accelerated with specialized tensor cores. This replacement can be hidden to the framework end-user, where everything seems to run in *fp*32. The *fp*32 to *fp*19 conversions (truncating the last 13 mantissa bits) and the *fp*19 MACs are managed by the CUDA compiler and hidden by low-level libraries, such as cuDNN and cuBLAS. The accuracy of *fp*19 MACs is not guaranteed to be the same as *fp*32 MACs. However, empirical evidence using *bf*16 (which carries to *fp*19) suggests that for DL workloads, the accuracy difference is insignificant; although unknown outliers may exist [KMM+19].

The primary advantage of *tf*32 is the ease-of-adoption. It requires no changes in the DL libraries (except for an enablement flag) and works out-of-the-box. The disadvantage is the lack of memory or bandwidth savings compared to 16-bit formats, which is often the bigger bottleneck.

Integer-16 (*int16*) training has been demonstrated on some models with no hyperparameters tuning [KWW+17, DMM+18]. The distribution of the weights, activations, weight gradients, and activation gradients in a tensor can be represented using *int16* and one shared scalar for the entire tensor. This scalar is dynamically adjusted to maximize range and minimize overflow. The weight and activation distributions do not change rapidly in consecutive training iterations. The gradient distribution changes more rapidly. A program can monitor the distributions and adjust the exponents for each tensor as needed.

For training, *int16* is not used in production; *bf*16 and *fp*16 are preferred over *int16* given the added complexity to manage the shared exponent with *int16*, particularly for the gradient tensors. For inference, *int16* has some adoption. Habana Goya uses *int16* for workloads that required more precision than *int8* (Habana Goya also supports other formats) [Hab19].

Integer-8 (*int8*) is rapidly gaining adoption for *some* inference workloads. Using *int8* often reduces the statistical performance due to the information loss quantizing from 32-bit to 8-bit. For some applications, a small drop in statistical performance is unacceptable, as it can have a negative monetary impact. In particular, less relevant product recommendation results in reduced purchases. There are techniques to reduce the statistical loss discussed in Section 6.2. Note that training with *int8* is limited to academic research on a few simple models not relevant in industry.

There are two main challenges with most *int8* quantization techniques. First, the uniform distribution of *int8* does not allow finer-granularity to better represent values in high-density regions where most of the information exists. A better approach is to use a nonuniform numerical format with high granularity in high-density regions and low granularity in low-density regions. This reduces the 32- to 8-bit information loss. Some proposals, such as *fp*8, are discussed below.

Second, precomputing the activations' quantization factors is needed to maximize the computational benefits of *int8* but requires additional effort for the developer. The distribution of the activation values with production data can be estimated using data samples with similar characteristics as the production data. This requires that a developer quantizing a model has access to production-like data samples.

Despite these challenges, *int8* is supported by all prevalent hardware marketed for inference. Google uses *int8* in production on TPUs for some MLP-, CNN-, and LSTM-based models, and on the Google Pixel phone for speech recognition with RNN models. Facebook (as well as many other companies) also uses *int8* across various workloads [JYP+17, HSP+19, PNB+18]. Facebook also demonstrated quantization to 4 bits on the embedding layers for serving recommendations without affecting statistical performance.

In particular *int8* inference has been shown to work across various CNN models [GMY+19]. However, even some CNN models like MobileNet and ResNeXt, and various non-CNNs such as BERT, are more susceptible to information loss from quantization and require additional effort to achieve acceptable statistical performance [SDY+19]. While the acceptable degradation varies, for most companies degradation over 1% is unacceptable, under 0.5% is acceptable, and in between depends on the application. Recommenders have a stricter threshold in the order of 0.01% due to the monetization impact.

Floating-point 8-bit (*fp8*) is used by Microsoft in FPGAs (Microsoft also uses *fp9*) using either 2 or 3 mantissa bits. *fp8* is implemented by researchers in some ASICs, such as the deep-learning neural processing unit (LNPU) to demonstrate *training* models on mobile devices (LNPU uses *fp8* and *fp16* mixed precision training) [CFO+18, LLH+19]. Intel and IBM demonstrate that *fp8* multiplies (accumulated to *fp32* and *fp16*, respectively) can be used for training and inference with insignificant loss in performance for various workloads [CBG+20, MSD+19, SCC+19].

There is no standardized *fp8* format. The most common formats are $(1, 5, 2)$ and $(1, 4, 3)$. The $(1, 5, 2)$ format better represents the dynamic range of the gradients. A particular challenge in training with an 8-bit format is in RNNs and models without normalization layers, as they are more susceptible to errors. The gradient errors can quickly increase in RNNs, and the typical lack of normalization can result in irregular tensor value distributions.

IBM proposed a hybrid $(1, 4, 3)$ and $(1, 5, 2)$ approach for the forward and backpropagation, respectively, using loss-scaling and stochastic rounding, and keeping the input and last layers at *fp16* [SCC+19]. The $(1, 4, 3)$ format is modified using a -4 fixed exponent bias to shift the coverage range by 2^{-4} to better align with the distribution of the weights and activations. This format is referred to as *fp8-ibm* in Table 6.1. There are two primary challenges to this format. First, some models, such as GNMT and Transfomer, require dynamic loss to properly converge, which increases the software complexity. Second, the more limited representation of small values, compared to *fp16* (the smallest positive values are 1.5×10^{-5} in $(1, 5, 2)$ vs. 6.0×10^{-8} in $(1, 5, 10)$, often results in underflow.

Intel has proposed two methods, both using the $(1, 5, 2)$ format. One method uses a shift and scale (shifted and squeezed FP8 (S2FP8)) parameter per tensor to represent a broad set of values. S2FP8 alleviates the need for loss-scaling, stochastic rounding, and $fp32$ for the first and last layer. The main weights and accumulations are in $fp32$ [CBG+20]. However, S2FP8 requires tracking the statistics in the tensor distribution (similar to $int16$ training) and updating the shift and scale parameters which increases the software complexity.

The other method uses enhanced loss scaling to improve the range of values and reduce the common underflow observed with $fp8$ training. This method uses loss scaling with a dynamically increasing minimum threshold for the scaling factor. Using a minimum threshold ignores spurious overflows in order to maintain a higher loss scale value. However, this method requires observing the training cost to determine when to adjust this threshold value.

A significant advantage of $fp8$ over $int8$ inference is circumventing the complexities of quantization. The current disadvantage is the limited hardware and software supporting $fp8$ formats. A minor disadvantage is that NaNs are overrepresented and consume 6 out of 256 (2%) and 14 out of 256 (6%) values in the $(1, 5, 2)$ and $(1, 4, 3)$ formats, respectively.

The published $fp8$ empirical results suggest that for the backpropagation $(1, 5, 2)$ is preferred over $(1, 4, 3)$. For inference (forward propagation), IBM demonstrated superior statistical performance using $(1, 4, 3)$ with the exponent shift, albeit the results are primarily targeting convolutional models. Intel demonstrated $(1, 5, 2)$ for both forward and backpropagation across ResNet, GNMT, Transformer, and NCF. The published results suggest that CNN models can benefit more from the additional mantissa bit in $(1, 4, 3)$, and non-CNN models can benefit more from the additional exponent bit in $(1, 5, 2)$. Nevertheless, the number of models in these studies is relatively small, and making solid conclusions requires further work.

Integer-4 ($int4$) support is available in recent Nvidia GPUs. $int4$ inference adoption on some CNN models may slowly grow on edge devices, such as in mobile phones, where power and memory are limited. The adoption in data centers may likely be none to very limited for workloads tolerant to extremely low range and precision and limited to representing activations from ReLU functions with unsigned $int4$ (the weights kept at $int8$). There is ongoing research toward improving $int4$ quantization [CWV+18, Don19, GMY+19].

Floating-point 24-bit ($fp24$) $(1, 8, 15)$ is used by Alibaba Neural Processing Unit (NPU) for CNN models for the element-wise and reduction operators (the matrix-wise operators use $int8 \rightarrow int16$) [JHJ+20].

Posit is a relatively new format different from the IEEE floating standard. This format requires less power and die area than the IEEE floating-point counterpart [Gus17, Joh18]. It does not overrepresent NaNs and provides other benefits and drawbacks [dDF+19]. However, this format has minimal adoption in academia and none in industry.

Log-domain is another form of nonlinear quantization that has been shown to maintain statistical performance with smaller numerical formats [LMC+17]. This format has limited adoption in academia and none in industry.

Binary (1 bit) and **ternary** (2 bits to represent -1, 0, and 1) have been used in research, in particular, to represent the weights in a forward propagation passes [ROR+16, HS14].

Die Cost

The die cost to build a multiplier, and the power cost to use the multiplier both exhibit quadratic growth with the number of mantissa bits and increase linearly with the number of exponent bits. Therefore, a *bf*16 multiplier is less expensive than a *fp*16 multiplier. However, area costs continue to decrease rapidly, and therefore this difference should not be a major factor in the DL hardware design decisions. Usability and software development costs are much more critical factors.

To facilitate transitioning from hardware that only support one format (*fp*16 or *bf*16), we recommend designing hardware that supports both *bf*16 and *fp*16 formats using a 19-bit $(1, 8, 10)$ floating-point unit (FPU). Similarly, we recommend supporting both $(1, 5, 2)$ and $(1, 4, 3)$ *fp*8 formats using a 9-bit $(1, 5, 3)$ FPU. According to IBM, supporting both formats only requires a 5% larger unit than supporting one format [SCC+19].

6.2 QUANTIZATION METHODOLOGY

Using *int8* can improve the computational performance at the expense of some (1) additional development and (2) loss in statistical performance. In this section, we explain the quantization methodology and share techniques that can mitigate loss in statistical performance and reduce the development process.

Assuming an *fp*32, *fp*16, or *bf*16, trained model a simple technique to quantize to *int8* is as follows: For each weight tensor, the maximum absolute value is mapped to ±127. For the activation tensors, a representative sample of the production data, called the *calibration dataset*, is used to collect activations statistics to find the distribution of activation values in each tensor across the samples. The quantization factor is:

$$Q_{\mathrm{a,w}} = \frac{127}{\max(abs(T_{\mathrm{a,w}}))},$$

where $T_{\mathrm{a,w}}$ is a tensor corresponding to either the weights $\mathbf{w}$ or the activations $\mathbf{a}$ (recall that the inputs to the NN can be considered the activations of Layer 0). The quantized values are:

$$\mathbf{a}_{s8} = \Phi\left(Q_{\mathrm{a}}\mathbf{a}_{f32}\right) \in [-127, 127]$$
$$\mathbf{w}_{s8} = \Phi\left(Q_{\mathrm{w}}\mathbf{w}_{f32}\right) \in [-127, 127],$$

where the function $\Phi(\cdot)$ rounds to the nearest integer.

The following techniques can improve *int8* inference accuracy. Note that even with these techniques, the loss over *fp*32 accuracy may still be unacceptable for some applications.

Asymmetric quantization uses a scalar and a shift factor, which can improve the quantization of the *activations*. Note that the weights are typically approximately zero-mean and should use symmetric quantization. The minimum activation value gets mapped to -128 and the maximum value to 127.

Threshold calibration requires deployment-like data (unlabeled data is OK) and no additional backpropagation. Mapping the largest absolute value to ±127 (or in asymmetric quantization the minimum and maximum value to -128 and 127, respectively) may result in poor utilization of the available 256 *int8* values when an outlier number is much larger than the other numbers. To illustrate, suppose the largest number is $10\times$ larger than the next largest value. That one number gets mapped to 127, and the rest of the values can only map to $[-13, 13]$. It is better to ignore outliers and find a threshold that minimizes the reconstruction error back to *fp32*. Another approach that works for some CNN models is to truncate outliers to minimize the information loss measured by the KL-divergence between the larger numerical representation tensor distribution and the quantized tensor distribution [Mig17]. Note that KL-divergence minimizes a metric of error in a layer, which may not minimize the accuracy error in the entire model. In practice, just using a threshold that captures 99% or 99.9% of the values results in superior performance accuracy.

Quantization aware training (QAT) requires labeled data (training data) and backpropagation. QAT (as opposed to post-training quantization) fine-tunes a model while enforcing quantization, and has been shown to improve accuracy. At each training iteration, the weights and activations of the layers targeted for quantization are fake-quantized to mimic *int8* values. The cost used in the backpropagation is based on the quantized values. The gradients and the weights updates are computed in single-precision. Another advantage is that QAT eliminates the need for the threshold calibration step as QAT minimizes the reconstruction error of the quantized values.

Selective quantization requires labeled data but no backpropagation. Some layers, such as softmax, tanh, sigmoid, depthwise-separable convolution, GELU, and the input and output layers, are more sensitive to quantization and should be kept at the larger numerical format to reduce the accuracy loss [Wu19]. The sensitivity of softmax can be slightly reduced by accumulating the logits in the larger numerical format and subtracting the max value before quantizing [BHH20]. The activation output of GELU can be clipped, for instance, to 10, in order to allow some *int8* value to represent the GELU negative activation values.

Analyzing an approximation of the Hessian matrix's trace is recommended to assess the sensitivity of a layer. This technique can be used to reduce the numerical format to 4 bits for some layers with minimal accuracy loss [DYC+19]. Other less reliable but faster-to-compute metrics to assess sensitivity are the KL-divergence, and the root mean squared error (RMSE) with the reconstructed *fp32* model. RL can facilitate designing a quantized model optimized for latency, energy, and accuracy for a particular hardware target. A possible algorithm for selective quantization follows:

Algorithm 6.2 Quantization Technique

Quantize all the layers and approximate the average Hessian trace for each layer [DYC+19]

Set the maximum acceptable accuracy error E

while *accuracy error* $> E$ **do**

    Unquantize the 8-bit (or 4-bit) layer with the highest average Hessian trace

---

This algorithm determines the layers that can be quantized. Note that one challenge is that interleaving layers with large and small numerical formats may result in higher computational cost from the overhead of the many conversions.

Cross-layer range equalization is a data-free quantization (requires no data and no back-propagation). The range of weights across the layers is equalized, and the range of activations are constraint under the assumption that a piece-wise linear activation function (such as ReLU) is used between the layers [NvB+19]. This constraint is satisfied by many CNN models but not by non-CNN models. This technique is used in the Qualcomm Neural Processing SDK.

Channel-wise quantization uses a quantization factor for each channel rather than one factor for the entire tensor.

Stochastic rounding (rather than nearest-value rounding) after multiplying by the quantization factor can improve performance [WCB+18]. To illustrate, rather than rounding the number 1.2 to the number 1, it is rounded to 1 with 80% probability and to 2 with 20% probability.

Unsigned *int8* **ReLU activations** uses the unsigned *int8* representation, rather than signed *int8*, for the activations of the ReLU functions. Using signed *int8* wastes half of the values since all the activations are nonnegative.

The techniques QAT, selective quantization, channel-wise quantization, and stochastic rounding also benefit *fp8* [CBG+20].

6.3   PRUNING AND COMPRESSION

Trained models typically have several weights that are approximately zero. Pruning them, that is, forcing all the weights less than some small ϵ value to zero results in a sparse model. Selecting a good value for ϵ requires experimentation. Pruning has been used for several decades to reduce the size of models. An interesting (but likely just coincidental) side note is that pruning biological neurons is important for healthy development [Iva71, LDS89, JS18, Wal13]. While pruning can reduce the number of operations using sparse operators, the primary benefit of pruning is to reduce the memory footprint via compression and alleviate memory bandwidth constraints. Note that AutoML, discussed in Section 10.1, can be used to learn a compact topology [HLL+19].

Doing some pruning usually has minimal impact on statistical performance, depending on the amount of pruning. In some cases, it may improve performance as pruning is a form of regularization. The ability to prune a model without affecting the statistical performance means

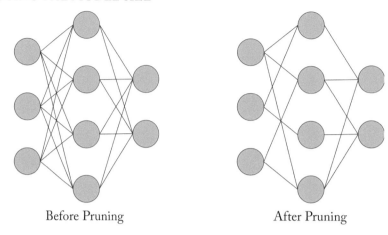

Before Pruning   After Pruning

Figure 6.2: Pruning a model by removing the weights (links) closed to zero.

the model is overparameterized. A hypothesis is that overparameterized models are needed to better explore the solution space and find a flatter minimum. After training the model, many of those parameters are no longer needed. A related hypothesis is the Lottery Ticket: within a large model there exist smaller models (lottery winners) that have the same or better performance as the larger model [FC19].

There are two types of model sparsity: structured and unstructured. Structured sparsity learning (SSL) prunes an entire vector, array, or tensor. SSL reduces the overall number of parameters and computations; for instance, by removing a convolutional filter [MHP+17]. Various SSL techniques have been developed [WWW+16, HGD+17, ZTZ+18, ZDH19, HZS+19, LSZ+19]. On CPUs and GPUs, structured sparsity (unlike unstructured sparsity) can reduce the number of operations.

Unstructured sparsity prunes values throughout a tensor without affecting the overall structure of the tensor, as shown in Figure 6.2. The unstructured sparse pruned model can take advantage of BLAS functions in the Nvidia cuSPARSE and Intel oneMKL libraries when the sparsity is greater than 90%. However, most sparse models have insufficient sparsity to significantly benefit from the sparse GEMM functions in these libraries. Alternatively, Google, DeepMind, and Stanford developed techniques that achieve $1.2 \times -2.1\times$ speedups and up to $12.8\times$ memory savings on Nvidia V100 GPUs without sacrificing accuracy on moderately sparse Transformer and MobileNet models [GZY+20].

Most production hardware are designed for dense matrix operations. Hardware with support for sparse operands is limited; one example is the LNPU device [LLH+19]. Nvidia A100 GPUs have support for fine-grained structure sparsity with $2\times$ more compute.

The techniques for pruning are:

- train with larger weight decay to force more weights near zero;

- fine-tune the pruned model (requires labeled data and backpropagation) [HPN+17]; and

- prune throughout the training process: set the small weights to zero at each training iteration [LCZ+19].

For power-constrained edge devices, energy aware pruning may be required; that is, pruning the layers that consume the most energy [YCS17].

Pruned models are less robust to adversarial attacks. An adversarial attack occurs when the input to the NN is meticulously altered so that a human would not detect the change, but the model produces a very different output. For instance, the model predicts with high confidence that the imperceivable altered image of a bus is an ostrich. Adversarially Trained Model Compression (ATMC) and Defensive Quantization are techniques that provide a balance between pruning and ensuring robustness to these attacks [GWY+19, LGH19].

Model compression reduces memory and bandwidth requirements at the expense of some additional computations for decompression. The time for these additional computations is often small relative to the time saved from the reduced bandwidth constraints. Therefore, compressing is usually advantageous. Note that an uncompressed unstructured sparse model and a dense model have the same memory footprint because storing an uncompressed zero-value requires the same number of bits as any other value. Compression algorithms, such as Huffman coding, use 1 bit to encode common values, such as the zero value. Another technique is to cluster similar values and to quantize them to few bits, each group having a quantization factor [HKK16].

Models with ReLU functions have sparse activations, and that sparsity grows for activations deeper into the model. During the forward propagation training stage, compressing the sparse activations before storing them (to use for the backpropagation stage) alleviates bandwidth bottlenecks.

6.4   KNOWLEDGE DISTILLATION

Knowledge distillation (KD) is a model compression technique that builds on the work by Bucila et al. and is gaining rapid adoption [HVD15, BCN06]. KD reduces the memory and computational requirements for a particular task and does not require a decompression step. KD is related to transfer learning. The knowledge from a complex model (the teacher model) is distilled to a simpler model (the student model). The student model is trained using a smaller dataset and a larger LR than was used on the teacher model.

The trained teacher model generates softened probability outputs on the student's training dataset. The student model is trained to produce similar outputs as the teacher's softened probability output, as illustrated in Figure 6.3. A softened softmax, also called a softmax temperature, first divides the logits by some value $T > 1$ (called the temperature) before normalizing them. The output is a softened probability distribution that better captures class similarities. To illustrate, the softened output in digit classification for an input image with the number 7 should

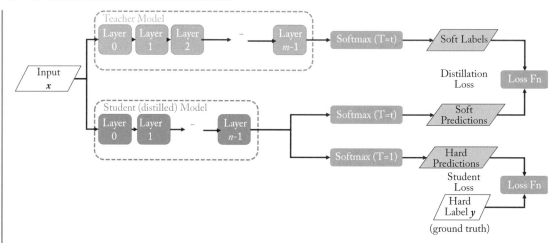

Figure 6.3: Knowledge distillation. A large teacher model distills the knowledge to a smaller student model. The student model learns using both the regular softmax and a softened softmax from the teacher model. Based on [Int18].

have the highest value for 7 and also a relatively high value for digits that look like 7, such as the handwritten digit 1 and 9. The student model is trained to learn (1) the softened output using a softmax temperature and (2) the one-hot ground truth vector using the regular softmax. The softmax temperature also provides regularization to the model [YTL+19].

The intuition behind KD is that the teacher model requires a more complex model to learn the relationships between the various classes. The ground truth one-hot vector does not encode class similarities and treats each class as entirely independent. The teacher model provides the class relations to the student model. Thus, the student model does not need to learn them from scratch and can use a simpler topology.

Extensions to this work are the deep mutual learning (DML) where an ensemble of students collaboratively learn and teach others by sharing their softmax outputs, and the teacher assistant (TA) to distill the knowledge from the larger-size teacher model to an intermediate-size TA model to a smaller-size student model [ZXH+17, MFL+19].

In this chapter, we detailed the various numerical formats used in production and those in exploration by researchers as well as compression techniques to reduce the memory footprint of models. Using a smaller numerical representation can increase the number of operations per cycle, and reduce the memory, memory bandwidth, network bandwidth, and power consumption. However, it may also result in lower statistical performance, particularly for some *int8* models. We discussed advances in quantization techniques to mitigate this accuracy loss and find Hessian-based analysis as a promising path to determine which layers are quantizable. Hardware support across numerical formats is one of the vital hardware design decisions. We recommend that training processors primarily support both *bf*16 and *fp*16 given the small die cost over supporting just one, and some *fp*32, and inference processors primarily support *fp*16, *bf*16 for compatibility with the training format, *int8* and *fp*8 and some *fp*32. In the next chapter, we review the basics of computer architecture, and discuss the various DL hardware designs.

CHAPTER 7

Hardware

The primary components in a DL platform are multitudinous multiplication and addition units, sufficient memory capacity, high memory bandwidth to feed the compute units, high inter-node and inter-server bandwidth for distributed computing, and power to operate. The tradeoffs of architecting DL hardware depend on the targeted workloads and operating environment. The enormous design space includes numerical formats, memory hierarchies, power constraints, area constraints, software- or hardware-managed caches/scratchpads, support for dense and sparse computations, domain-specific to general-purpose compute ratios, compute-to-bandwidth ratios, inter-chip and inter-server interconnects, and ease of programmability.

The cost of arithmetic logic units (ALUs) is decreasing, and computational capacity is growing faster than memory bandwidth, as shown in Figure 7.1 for the top supercomputer. The primary hardware bottlenecks executing DL workloads are:

- main memory bandwidth;

- local (SRAM) memory; and

- power (primarily from data movement).

Moore's Law continues to deliver exponential growth in the number of transistors that can be packed into a given area, albeit at a slower rate than before. Computer architects are finding new ways to extract performance from this exponential growth. However, as a consequence of this exponential growth, compute and memory capacity are increasing much faster than memory bandwidth, which is the bottleneck in many DL workloads. The slow growth in bandwidth relative to compute is known as the memory wall or bandwidth wall, where compute units are idled waiting for data [WM95, RKB+09].

As transistors shrink, their power density no longer stays constant but rather increases, which is known as the end of Dennard's scaling (discussed in Section 7.1) [DGY+74]. The amount of dark silicon, where transistors cannot operate at the nominal voltage, is increasing. This dark silicon motivates the exploitation of transistors for multicore processors and domain-specific circuitry. Some of the existing techniques to increase performance are (detailed in Section 7.4):

- using a memory hierarchy to facilitate data-reuse;

- increasing the memory bandwidth;

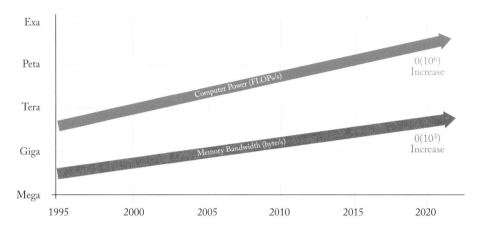

Figure 7.1: Computational capacity is growing faster than memory bandwidth as measured by the capacity of the top supercomputer. Based on [LHL+18].

- placing the memory close to the compute units to reduce access time and energy;

- applying a single instruction to multiple data;

- reducing the numerical representation and compressing the data; and

- using specialized logic or dedicated accelerators.

Each innovation is a one-time card; that is, each innovation gives a performance boost once since these innovations do not resolve Dennard's scaling. From a clock-speed perspective, devices are not getting any faster.

In order of decreasing access time and increasing die area, the storage types are: nonvolatile memory (flash memory, magnetic disk), DRAM (HBM2/E, GDDR6, DDR4, LPDDR4/5), SRAM (scratchpad, cache), and registers, all discussed in Section 7.2. DRAM is often called main memory and SRAM local memory.

Table 7.1 compares the energy for various operators, including data fetching, in a von Neumann architecture. DRAM access can cost two orders of magnitude more power and processing cycles over local SRAM. It is crucial to minimize DRAM accesses to improve performance and reduce power consumption.

A smaller numerical representation compute unit requires less area and energy; thus, more units can be added to a given die. The amount of area needed for floating-point multipliers grows quadratically with the number of mantissa bits and linearly with the number of exponent bits.

In Section 1.8, we introduced (1) the high-level requirements for the different DL usages: hardware design, topology design, training in production, serving in data centers, and serving in edge devices; and (2) the important hardware features and MAC combinations for training and

Table 7.1: Required energy and silicon area for various operator units in a 45 nm technology node. Adopted from [Hor14] and [Dal16].

Operator:	Energy (pJ)	Area (μm^2)
8b Add	0.03	36
16b Add	0.05	67
32b Add	0.1	137
16b FP Add	0.4	1,360
32b FP Add	0.9	4,184
8b Mult	0.2	282
32b Mult	3.1	3,495
16b FPMult	1.1	1,640
32b FP Mult	3.7	7,700
32b SRAM Read (8 KB)	5	N/A
32b DRAM Read	640	N/A

serving production hardware. Training requires storing and retrieving the activations across all the layers, which typically involves reading and writing several GB of data (the activations) from and to DRAM. In training CNNs, the size of the activations typically has a more significant impact on the total memory requirements than the size of the model. To illustrate, U-Net (used for medical 3D image classification) has 20 million weights but requires 256 GB of memory. Conversely, Megatron-LM-1.2B has 1.2 billion weights but requires 32 GB of memory. Given the amount of data transfer, using a high bandwidth DRAM, such as HBM2E, for training tasks is beneficial. An advantageous design choice is to put enough SRAM to store the model and the activations associated with two consecutive layers in training and inference. Note that the size of the activations is proportional to the batch size, which is usually small for inference.

As much as possible, data center managers want a homogeneous and manageable data center leveraging specialized accelerators only when absolutely needed. However, given the exponential demand for compute and the end of Dennard's scaling, the demand for dedicated DL processors is increasing. Hardware designers should be aware of what hyperscalers value:

1. performance per power and per cost;

2. minimal transition work from an existing to a new hardware/software system;

3. programmability and ease-of-use; and

4. high utilization (the device does not sit idle frequently).

Flexibility is a crucial consideration when designing DL hardware, in particular, because the design is mostly locked around two to three years before the product is deployed, which is a challenge in this rapidly evolving field. Recent algorithmic advancements include depthwise separable convolution, dilated convolutions, residual connections, ReLU variants, and GNNs. New models have irregular memory access patterns, more complex control-flow, and dynamic computation graphs that vary with different input data and cannot be optimized at compile time. These models can benefit from higher general-purpose compute. Models with predictable access patterns and primarily dense linear algebra benefit from dedicated matrix multipliers. Note that the required compute and bandwidth can significantly vary for different workloads, as shown in Figure 1.16.

The following are important hardware features for training and serving hardware for production workloads in data centers. Software is equally important and discussed in Chapters 8 and 9. The hardware characteristics for training are:

- masses of $bf16 \rightarrow fp32$, $fp16 \rightarrow fp32$, and sufficient $fp32 \rightarrow fp32$ MACs;

- high in-die interconnect (IDI) bandwidth for multicore GEMMs and broadcast/reduce collectives;

- sufficiently large SRAM for the weights and some activations;

- high DRAM bandwidth to store and read the activations, or alternatively, a much larger SRAM capacity;

- high intra-server inter-node bandwidth for (1) multinode GEMM, (2) broadcast/reduce collectives in large embedding layers, and (3) distributed training across nodes (on servers without accelerators, a node can be a socket or a group of cores); and

- high inter-server bandwidth for distributed training.

For serving:

- plenty of $fp16 \rightarrow fp32$, $bf16 \rightarrow fp32$, $int8 \rightarrow int32$, $fp8 \rightarrow fp32$ and some $fp32 \rightarrow fp32$ MACs;

- high IDI bandwidth for multicore communication for GEMMs and broadcast/reduce collectives;

- sufficiently large SRAM for the weights and some activations; and

- video decoder for media analytic workloads (inference on video frames).

Established hardware companies and several startups have developed or are developing specialized processors and platforms with CPUs, GPUs, FPGAs, CGRAs, DSPs, ASICs, and a mixture of them. The number of DL processors is likely to expand over the next few years and later consolidate. Most of the processors are not going to succeed in the market for two reasons. First, multiple highly specialized processors result in limited volume for each processor, which limits the economies of scale and may limit access to the most advanced fabrication processes. The market demand is further limited as some hyperscalers requiring a large volume of processors have built or are building their own. Second, DL engineers are unlikely to learn multiple programming paradigms. The companies that succeed are likely those that delivered a cost-effective programmable and balanced platform with some flexibility to adjust for different requirements, such as the general to domain-specific compute ratio.

In the remainder of this chapter, we review the basic concepts of computer architecture and hardware designs, discuss the various types of memories, and explain workload roofline analysis. Lastly, we compare various DL processors and platforms and their strengths and weaknesses.

7.1   MOORE, DENNARD, AND AMDAHL

Demand for more compute, memory, and bandwidth is growing at a time when exponential compute growth is slowing down. In this section, we review the reasons for this slowing down and the ways hardware vendors are partially mitigating this.

At the heart of compute and memory is silicon-made transistor gates. Transistor gates form logic gates, such as the AND, OR, and NOT, which in turn form more complex logic. Transistor gates open and close to block (insulate) or let (conduct) electrical current through by charging or discharging the gate capacitance using voltage. The capacitance is related to the length and width of the transistors. On the one hand, the smaller the capacitance, the less voltage the transistor needs, and the faster it can operate. On the other hand, the less voltage, the less tight the gate closes, and the more current it leaks. Also, as transistors get smaller, both the length of the channel (the region between the source and drain under the gate) and the insulation layers are reduced, which increases the electron tunneling between the source and drain. This tunneling results in current leakage and, as discussed below, is the primary reason for the slowdown in computational growth.

Gordon Moore predicted in 1975 that for the next several years the number of transistors that fit in the same chip area would double every two years (revising his 1965 prediction of doubling per year) through improved technology and innovative techniques [Moo75, Moo65]. This observation is known as *Moore's Law* and as held for several decades. Today, the exponential growth continues, albeit with a longer than two-year cadence. The primary benefits are less expensive processors, more powerful processors with more transistors and logic per area, or both.

Robert Dennard et al. demonstrated that as transistors shrink, their power density stays approximately constant [DGY+74]. This is known as *Dennard's scaling*. Combined with Moore's Law, every two years and under the same power budget, the number of transistors per area doubled and operated at 40% higher frequencies (details below). Dennard's scaling broke down in the mid-2000s due to current leaking.

The total consumed power (in Watts or Joules per second) is the sum of the dynamic (or switching) power and the static (or leakage) power. Dennard scaling only accounts for the dynamic power, which is defined as follows:

$$P_D = Q \cdot E \cdot f = \frac{1}{2} \cdot Q \cdot C \cdot V^2 \cdot f,$$

where $E = \frac{1}{2} \cdot C \cdot V^2$ is the energy (in Joules) to open or close a transistor gate, Q is the number of active transistors (here we assume all transistors are active), C is the capacitance, V is the voltage, and f is the frequency. Scaling *down* the transistor dimensions by $\kappa > 1$, scales down the area by κ^2, and scales down the capacitance, charge/discharge time (inverse of frequency), and voltage by κ. The number of active transistors under the same area is κ^2, and the dynamic power remains constant:

$$P_D = \frac{1}{2}(Q \cdot \kappa^2)(C/\kappa)(V/\kappa)^2(f \cdot \kappa) = \frac{1}{2} \cdot Q \cdot C \cdot V^2 \cdot f.$$

To illustrate, if the transistor dimensions scale by $\kappa = \sqrt{2} \approx 1.4$ (per Moore's prediction), then under the same dynamic power and area, the number of transistors doubles and the frequency increases by 40%.

The static power is:

$$P_S = V \cdot I_{leakage},$$

where $I_{leakage}$ is the current leakage. Dennard's scaling ended in the mid-2000s due to the increases in static power. Today, reducing the voltage in smaller transistors increases current leakage and increases the power density. Instead of having more clock cycles per second (higher frequency), the focus today is on increasing the instructions per cycle (IPC) or operations per cycle; that is, doing more work per cycle.

Decreasing the voltage increases the *static* power exponentially. Increasing the voltage increases the *dynamic* power exponentially, as shown in Figure 7.2. There is an ideal voltage that minimizes the sum of the static and dynamic power.

The propagation time T_{prop} of the current through all the logic gates in its path needs to be less than 1 clock cycle. As the frequency increases past some $f_{\min}$, higher voltage is required to operate the transistor gates more rapidly. This increase in voltage is approximately linearly proportional to the increase in frequency, as shown in Figure 7.3. The voltage to operate at $f_{\min}$ is $V_{\min}$. Increasing the frequency past $f_{\min}$ increases the power to the cube of the frequency increase: $\Delta P \propto (\Delta V)^2 \Delta f \propto (\Delta f)^3$.

Power generates heat, and too much heat can damage the circuits. There is a maximum power that a system can operate without damaging the circuitry, and this limits the maximum frequency. In fact, over the past decade, the maximum frequency of high-end server processors has not changed much. Servers continue to operate in the 2–4 GHz range. Another reason to cap the maximum frequency is related to the maximum distance the electrical current travels in

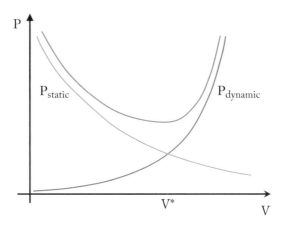

Figure 7.2: Total power requirement (red curve) across various voltages. Low voltage results in high static power due to current leakage. High voltage results in high dynamic power. There is an optimal voltage V^* where the total power usage is minimized.

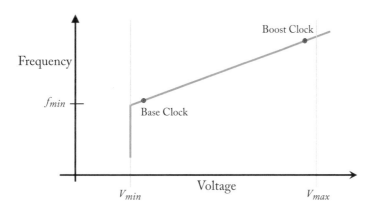

Figure 7.3: Increasing the frequency past f_{min} linearly increases the required voltage, and (not shown) cubically increases the dynamic power.

the circuitry. The time it takes to travel this distance must be less than one clock tick, which can be an issue for large circuits when operating at high frequencies.

The primary contributors to the increased dark silicon are the exponential growth in transistors per area, current leakage, and power constraints. Multicore processors and specialized computing are two methods to mitigate dark silicon. These methods have enable the continued growth in computational capacity at the expense of two new challenges: Amdahl's law and the memory wall.

Gene Amdahl formalized the speedup when only a fraction of a program is improved, known as Amdahl's law. It is used to determine the limitations of parallel computing [Amd67]. Using $N > 1$ cores for a particular workload results in a maximum speed up of

$$\frac{1}{(1 - P) + (P/N)},$$

where P is the percentage of the workload that is parallelizable. Approaching this maximum speed up requires nontrivial *parallel programming*, and there is a computer science field dedicated to this. Even assuming $P = 1$, perfect linear scaling across general-purpose multicores is not possible. There are core-to-core bandwidth limitations and cache coherence overhead, which grows with more cores.

These limitations and overheads are motivations to reduce the scope of hardware-based cache coherence and to use domain-specific DL processors for embarrassingly parallel (minimal communication/synchronization between parallel elements) workloads with predictable operations. Solutions still require a way to operate on the right data, and this drives a combination of application-specific hardware and software-based "coherence" [AVG+15, TKT+16, ADC11].

Figure 7.4 provides a high-level view of the trends in microprocessors. The number of transistors per area continues to grow exponentially, and the number of logical cores is following that same growth path; new transistors are primarily used for additional cores. In the future, the growth in the number of cores may slow down, and more transistors utilized for domain-specific acceleration. While frequency has already plateaued, single-thread performance continues to increase due to better instruction pipeline, improved branch prediction, out-of-order execution, larger instruction vectors, and specialized execution units, resulting in more IPC.

7.2   MEMORY AND BANDWIDTH

The growth in memory bandwidth has been significantly slower than the growth in computational capacity. The time to read data from memory is often the main hindrance to performance. Moreover, given current trends, this divide between the compute (OPS) and the data transfer (bytes per second) is increasing, which can result in unutilized compute capacity. As an example, the Nvidia T4 card supports up to 130 *int8* TeraOPS (TOPS) with 320 GB/s bandwidth [Nvi20b]. To keep full utilization, the T4 processor must perform an average of 406 operations on every read byte. One way to reduce exposure to the memory bandwidth is to use a cache memory hierarchy that stores frequently or immediately accessed elements closer to the compute element. The efficiency of caches is dependent on the working set having spatial or temporal locality that can exploit these hierarchies. There is a rich literature on arranging matrices, known as *blocking* and *caching*, to fit various memory caches and achieve high data reuse [Gvd08, CWV+14, GAB+18, ZRW+18].

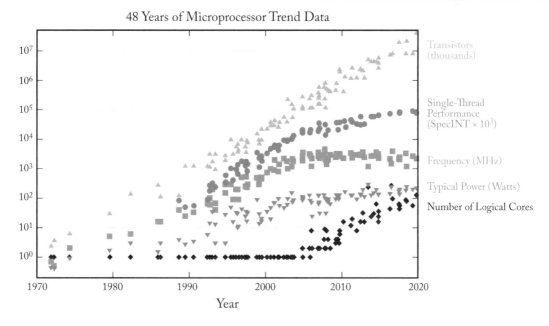

Figure 7.4: Trends in microprocessors. Source: [Rup20] (CC BY-SA 4.0 license).

Memory can be described by its capacity (bytes) and data transfer rate or bandwidth (bytes per second). The bandwidth (BW) can be computed as follows:

$$BW = f_{mem} \times \text{number of interfaces} \times \text{transfers per clock} \times \text{mem bus width},$$

where f_{mem} is the memory frequency, the interfaces are typically 2 (dual-channel configuration) in modern processors, and the transfers per clock are 2 for memories that transfer on both the rising and falling clock edge (such as DDR) and 1 otherwise. In practice, the effective transfers per clock may be slightly lower and workload-dependent; in DRAM, it depends on the distribution of read and write transactions.

The memory types used in production in increasing order of accessed time and, equivalently, in increasing order of memory density (bytes per silicon area) and decreasing monetary cost per byte are as follows:

1. processor registers;

2. SRAM: scratchpad, cache (typically with multiple levels); and

3. DRAM: HBM2/E, GDDR6, DDR4/5, LPDDR4/5.

There are two types of random-access memory: dynamic RAM (DRAM) and static RAM (SRAM). SRAM uses a bistable circuit design that is faster but more expensive and requires four to six transistors per bit. DRAM is slower but less expensive and requires only one transistor (and a capacitor) per bit, and hence it has higher memory density. The capacitor stores the charge (the bit). Reading the stored bit consumes this charge requiring a write after the read cycle to save the value. Even in the absence of read/write activity, DRAM memory must be frequently refreshed to avoid losing information as the charge leaks (at a temperature and device-dependent rate). This refresh involves reading the data and immediately writing it to the same area (as DRAM reads are destructive). SRAM does not require frequent reads and writes. Both DRAM and SRAM are volatile memories; that is, they lose the stored bits when the power is off.

There are two main types of SRAM configurations: caches and scratchpads [LAS+07]. A cache is implicitly addressed (not directly addressed by the software), hardware-managed memory. A scratchpad (also called streaming memory) is explicitly addressed, software-managed memory. Caches are common in CPUs and GPUs to support general-purpose workloads. Scratchpads are common in embedded and dedicated hardware, such as ASICs and DSPs, for static graph-based workloads to reduce power consumption.

A cache has additional logic circuitry to ensure cache coherence and improve locality to determine what data to keep (this data is known as *hot* entries or working set) and what data to replace. This logic alleviates the software (the programmer or compiler) from directly managing the cache memory access. However, it comes at the expense of higher energy cost per data access and lower memory density. This additional logic is beneficial for irregular access patterns, such as in GNNs, embedding layers, and DL dynamic graph-based models.

There can be different levels of caches. Modern CPUs have three-levels of caches: $L1$, $L2$ (mid-level cache (MLC)), and $L3$ (last-level cache (LLC)). $L1$ is the smallest and closest memory to the compute unit, and therefore has the fastest access time. CPU processors have two different $L1$ caches: a data cache unit (DCU or $L1d$) and an instruction cache unit (ICU or $L1i$). Data and instructions share the cache in $L2$ and $L3$. Modern GPUs have 2 levels of cache. The canonical chunk (block) of memory loaded from the main memory to the cache hierarchy is called a cache line. Note that loading an entire cache line can waste bandwidth and storage on sparsely strided memory accesses.

Different architectures use different cache replacement policy algorithms, and even different cache levels within an architecture may use different policies. While the specific policy used by a microarchitecture is not always made public, variants of the Least Recently Used (LRU) eviction policy are common, such as Adaptive Replacement Cache (ARC). LRU means the cache tracks and evicts the least recently accessed page when adding a new page. ARC tracks frequently used, recently used, and recently evicted pages.

While caches are hardware-managed, there is some work to enhance cache control with software hints. One example is using the CLDEMOTE instruction, which hints to the hardware to demote a given cache line to more distant cache from the processor to speed up access to the cache line by other cores (L1 caches are unique to a specific core).

A scratchpad has a simple memory structure that provides better efficiency at the expense of sophisticated software; it manages all the memory accesses and the replacement policy. A scratchpad is typically more efficient than a cache, usually 1–2 clock cycles per memory access. A scratchpad has addressable storage and requires explicit software-controlled direct memory access (DMA) transfers to orchestrate all data movement in the proper order. However, any mismatch of memory accesses to the ALU or FPU logic inputs or outputs may lead to orders of magnitude of performance degradation. Thus, scratchpads are typically limited to DL workloads with static graphs, where all data accesses are predictable and determined at compile-time. In high-volume production, saving some power and execution time has multiplicative benefits over the lifetime of the model, which may outweigh the software complexity costs.

A hybrid memory system uses both cache and scratchpad configurations. Nvidia architectures (excluding Pascal) configure some cache memory as a scratchpad for application-specific locality and communication optimizations. Note that Nvidia refers to scratchpad and cache as shared and automatic memory, respectively. There is research toward a unified configuration to get the best of both, such as Stash and Buffets [KSA+15, PSC+19].

There are three types of caches with different speeds and conflicts tradeoffs. Cache conflicts occur when a different cache line from memory maps to the same cache entry, thus evicting and replacing the existing cache entry. The placement depends on the memory address.

- *Fully Associative* places a cache line from memory in any entry in the cache; this has the slowest-access time but minimizes conflicts.

- *Direct Mapped* places a cache line from memory in a specific entry in the cache; this has the fastest-access time but maximizes conflicts.

- *N-way Set-Associative* places a cache line from memory in any of N entries in the cache; this provides a compromise between access time and conflicts.

In practice, most CPU caches in production are N-way set-associative caches. Understanding cache associativity can guide the design of the DL topology. To illustrate, an *fp32* GEMM with a leading dimension of 1024 (used in an RNN layer with 1024 units), results in high cache conflicts in CPUs; a better leading dimension is 1040 in modern CPUs, as explained in Section 7.2.1.

DRAM or, more precisely today, Synchronous DRAM, is less expensive in price and silicon area but is significantly more expensive in energy and access time compared to SRAM. There are various types of DRAM used in production: Double Data Rate (DDR), High-Bandwidth Memory (HBM), Graphics DDR (GDDR), and Low-power DDR (LPDDR), and various generations within each type [GLH+19]. DDR memories fetch the data on both the leading and falling edge of the clock signal. Other types of DRAM with minimal market adoption are Hybrid Memory Cube (HMC) and Wide I/O (WIO).

DDR   DDR4 is the most widely used DRAM. It is available in servers, workstations, laptops, and some inference accelerators, such as Habana Goya. Increasing the number of main mem-

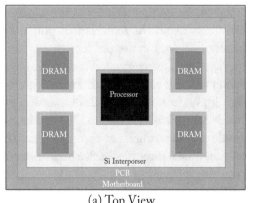

(a) Top View

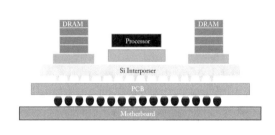

(b) Side View

Figure 7.5: HBM memory connected to the processor via an interposer. (a) Top view. (b) Side view. Based on [Sam16].

ory channels improves bandwidth and partially mitigates the memory wall [Hor14, PRH+17]. However, the maximum number of balls or pins possible on a package limits the number of channels. DDR5 is the latest generation of DDR providing higher bandwidth and density. Intel processors codenamed Sapphire Rapids and (likely) AMD processors codename Genoa should support DDR5.

HBM   HBM2 is the defacto DRAM memory for GPUs and accelerators targeting training, HPC, and cryptomining. It is available in the Nvidia {P, V, A}100 GPUs and Habana Gaudi. Google TPU v2 and v3 (and likely v4) use HBM but have not made public the specific HBM generation.

HBM2 has a 1024-bit wide interface across 8 channels per stack, and (in the latest specification) $\sim$2.4 GT/s transfer rates (each bus lane transfers $\sim$2.4 Gbps), for a total of 307 GB/s per DRAM stack or package. It provides higher bandwidth and uses less power relative to other DRAM memories. HBM memory connects to the processor via a purpose-built silicon chip called an *interposer* and mounts in the package substrate, as illustrated in Figure 7.5. The shorter wires allow for higher bandwidth at lower power. Given that HBM uses a stack of memory chips, it is referred to as 2.5D memory. An issue with HBM is the high price to manufacture the interposer, in part, because 2.5D is a relatively new memory technology. The cost may decrease as the technology gains broad adoption.

GDDR   GDDR6 is used in the latest gaming graphics cards and data center inference GPUs, such as the Nvidia T4, and may expand to other inference accelerators. Compared to HBM, GDDR is less expensive and has lower latency, but it also has lower bandwidth and lower memory density.

Set/Way 0 1 2 3 4 5 6 7

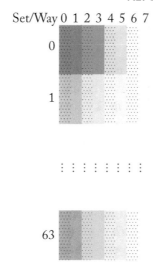

Figure 7.6: A representation of an 8-way set-associative cache with 64 sets.

LPDDR   LP-DDR4 and LP-DDR4X are widely used in low power devices, such as mobile phones. LPDDR has short wires and, therefore, low latency response. The newest generation LP-DDR5 is available in the latest mobile phones and expanding to other devices, such as tablets, ultra-thin notebooks, automotive, and tentatively, DL inference processors.

7.2.1   SET-ASSOCIATIVE CACHE

Understanding set-associative caches can guide the design of a DL topology. In an N-way set-associative cache, the cache is organized into multiple sets with N cache lines per set. Each location in the main memory maps to any of the N different cache lines of a given set. The index bits from the main memory address determines the set. Accessing a cache line in an already full set evicts one of the N entries already present. Note that direct-mapped cache can also be called a 1-way set-associative cache.

In an N-way associative cache, the number of sets is $S_N = sizeof(\text{cache})/(N \times sizeof(\text{cache line}))$ with $W_N = sizeof(\text{cache})/N$ as the interval of the main memory addresses that map to the same set. That is, equal addresses modulo W_N share the same cache set.

Figure 7.6 illustrates the cache lines in an 8-way set-associative $L1$ cache with $S_N = 64$ sets, common in modern CPUs. Each cache line is 64 bytes, the total cache size is $64 \times 64 \times 8 = 32KiB$ ($Ki = 2^{10}$), and the interval of the main memory addresses that map to the same set is $W_N = 32Ki/8 = 4096$ bytes. Note that in this discussion, $L1$ cache refers to $L1d$ (data) cache.

The analysis below assumes the matrices used in a program are organized in main memory in column-major order. That is, consecutive values along a matrix's column are consecutive in memory. Furthermore, the number of elements in each column (that is, the number of rows) is called the leading dimension of the matrix.

Data reuse is critical to increasing matrix-multiply (GEMM) performance. A common technique in a GEMM operator is to access the matrices' data in blocks that fit in the cache and to use those values for several computations before accessing another block. However, when a matrix with leading dimensions of 1024 $fp32$ values (4096 bytes) maps to the $L1$ cache, all the elements of a given row map to the same cache set, and after accessing 8 elements of a row, the set is full. A cache line is evicted to access another value in the row. Specifically, when the leading dimension is 1024 and a block of 16 rows by k columns is accessed (that is, $16k$ $fp32$ values), the whole block maps to only one set (16 $fp32$ values is 64 bytes or one cache line). If $k > 8$, the matrix block is evicted from cache every time it is accessed, which results in higher cache misses making caching less effective and reducing the GEMM performance. Using a leading dimension of 512 or 256, the $16 \times k$ block maps to 2 or 4 sets (out of 64), respectively, leading to slightly better but still poor GEMM performance.

For the $L1$ cache in this example, the best leading dimensions are 1040 and 1008 because the accessed matrix block spreads across the various cache sets. A recommendation for a data-scientist using $fp32$ is to use matrix sizes with a multiple of 16 (a cache line) as the leading dimension, but not a multiple of 256. This recommendation can influence the design of efficient DL topologies, for instance, designing RNN layers with 1008 units, rather than 1024.

More generally, for a given matrix, the leading dimension should be a multiple of the number of values that fit in a cache line, but not a multiple of $W_N/4$ and, in particular, not a multiple of W_N. GEMM software libraries, such as oneMKL, can partially mitigate poor leading dimensions by artificially modifying the matrices' sizes to more efficiently use the cache.

7.3   ROOFLINE MODELING

Roofline modeling estimates the maximum performance that a computational kernel or set of kernels can attain on a particular hardware [WWP09]. A simple roofline model has three components:

1. processor peak performance in operations (ops) per second (ops/s or OPS);

2. memory bandwidth in bytes per second (B/s); and

3. kernel arithmetic intensity (ops/B).

A processor's peak performance depends on the frequency, number of cores, number of ops per core per cycle, and hardware's efficiency. While some analyses use the theoretical peak, it is better to use the observed or estimated actual peak performance, which includes the processor's efficiency. This efficiency can be estimated based on historical data from highly optimized workloads, or by running a suite of micro-kernels, such as the CS Roofline Toolkit. This actual peak performance is processor-dependent and kernel independent. Similarly, running a suite of

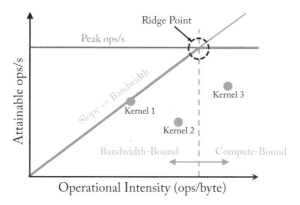

Figure 7.7: A roofline model models the maximum attainable OPS for a particular kernel on a particular hardware. Kernel 1 is well optimized and operating near the roofline. Kernels 2 and 3 are well below the roofline and require better software optimizations to more efficiently use the computational resources.

micro-kernels or an appropriate stream benchmark provides a more accurate observable bandwidth, which is a better metric than the theoretical maximum bandwidth.

The arithmetic intensity, also known as operational intensity (OI), is the ratio of the number of operations required to compute a kernel divided by the bytes read from DRAM memory. The literature usually uses *AI* to abbreviate arithmetic intensity, but we use *OI* to avoid confusion with *artificial intelligence*. The number of operations depends on the kernel and is typically independent of the processor. The number of bytes depends on both the kernel and the local SRAM memory size; a large SRAM facilitates data reuse.

A system with no SRAM is assumed to illustrate the worse case OI. If every operand is read from DRAM and every result is written to DRAM, then each operation (two reads and one write) would have an arithmetic intensity of $1/(3 \times sizeof(\text{datatype}))$. In the ideal case, the operands and result fit in SRAM and the OI is:

$$OI_{kernel} = \frac{ops}{sizeof(\text{input activations}) + sizeof(\text{weights}) + sizeof(\text{output activations})}.$$

In the literature, OI analyses sometimes assumes this best scenario, making OI independent of hardware. In practice, however, the OI depends on the memory hierarchy.

Figure 7.7 shows a roofline plot. The maximum attainable OPS for a kernel is the $min(bandwidth \times OI, peak\ OPS)$. Kernels where the attainable OPS are constrained by the $bandwidth \times OI$ are *bandwidth bound*, and those constrained by the *peak OPS* are *compute bound*. Increasing the computational capacity does not increase performance for bandwidth bound kernels. The relation between roofline and computation time is as follows: the time T it takes to execute a kernel, assuming perfect overlap of communication and computation, is:

$$T = \max\left(\frac{\text{number of ops to compute kernel}}{\text{peak processor OPS}}, \frac{\text{bytes to read from memory}}{\text{peak memory bandwidth}}\right).$$

Data reuse is key to achieving high OI. Data reuse means reusing the operands or the result for multiple cycles. The OI for a kernel function can vary considerably depending on how much data is reused. A traditional CNN kernel has high OI ($\sim$1000 ops/B), whereas a GEMM kernel used in an MLP, RNN, or other fully-connected layers typically has low OI ($\sim$10 ops/B) (see Figure 1.16).

The OI of a $\mathbf{C} = \mathbf{A} \times \mathbf{B}$ GEMM operation, assuming the data fits in SRAM, where $\mathbf{A} \in \mathfrak{R}^{M \times K}$, $\mathbf{B} \in \mathfrak{R}^{K \times N}$, and $\mathbf{C} \in \mathfrak{R}^{M \times N}$, is:

$$OI = \frac{2MKN}{sizeof(\text{datatype}) \times (2MN + MK + KN)},$$

where the 2 in the numerator is to account for multiplies and adds and the 2 in $2MN$ in the denominator is to account for reading and writing matrix $\mathbf{C}$ from and to main memory. A practical example is a fully-connected layer going from a layer with M units to a layer with K units and using a batch size of N and where matrix $\mathbf{A}$ is the weight matrix. Similarly, the OI of an $\mathbf{Z} = \mathbf{X} \otimes \mathbf{Y}$ convolution operation assuming the operands fits in SRAM, where $\mathbf{X} \in \mathfrak{R}^{NCHW}$, $\mathbf{Y} \in \mathfrak{R}^{KCRS}$, and $\mathbf{Z} \in \mathfrak{R}^{NK\tilde{H}\tilde{W}}$, is:

$$OI = \frac{2NKCRS\tilde{H}\tilde{W}}{sizeof(\text{datatype}) \times (2N\tilde{H}\tilde{W}K + KCRS + NHWC)}.$$

Element-wise operators have no data reuse and a very low OI. The OI can increase by fusing (merging) element-wise operators with computationally intensive operators, such as GEMM and convolution. For instance, the ReLU operator can be applied to the output of a convolution operation while the output data is still in the registers before writing it back to the cache or main memory.

Even when the operands do not fully fit in SRAM, GEMM and convolution operators can take advantage of data reuse. In the $\mathbf{C} = \mathbf{A} \times \mathbf{B}$ GEMM operation above, every value in matrix $\mathbf{B}$ is reused M times: every value in row $k \in [0, K-1]$ in matrix $\mathbf{B}$ is multiplied by all the M values in the corresponding column k in matrix $\mathbf{A}$. Every value in $\mathbf{C}$ is reused K times as it accumulates the K products. Weight reuse (the data in matrix $\mathbf{A}$) is proportional to the batch size N; a batch size of $N = 1$ has no weight reuse in a GEMM operation.

In the convolution operator, there is more data reuse. The weights of one filter $\mathbf{Y}_k \in \mathfrak{R}^{CRS}$ can be reused across the N dimension in the input tensor $\mathbf{X}$. Alternatively, the activations across one sample, $\mathbf{X}^{[n]} \in \mathfrak{R}^{HWC}$, can be reused across all weights $\mathbf{Y}$.

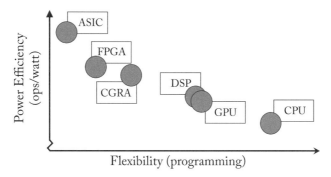

Figure 7.8: A generalization of different architectures providing tradeoffs between hardware efficiency and software flexibility (ease of programming). In practice, the actual tradeoff depends on the specific microarchitecture.

7.4   PROCESSOR DESIGNS

There are various types of architectures with different instruction sets, memory, and compute designs. In this section, we review instruction sets, and the type of processors used in DL, specifically CPUs, GPUs, FPGAs, CGRAs, DSPs, and ASICs, used separately or as components of a heterogeneous design. Given the prevalence of CPUs, GPUs, and ASICs, most of the focus is on these processors. These types of architectures balance flexibility (general-purpose computing) and performance (specialized computing), as illustrated in Figure 7.8 across generalized architectures. In practice, the actual tradeoff depends on the specific microarchitecture; for example, recent Nvidia microarchitectures have specialized (ASIC-like) compute units for matrix multiplications, which increases peak OPS at the expense of higher software complexity. In this section, we introduce key characteristics of each of these processors, and in Section 7.6, we discuss prominent processors in production and development.

The instruction set architecture (ISA) defines the operators, data types, and memory management for an abstract computer architecture. Different processors with different frequencies and memory sizes can implement the same ISA and execute the same binary. The specific implementation is called a *microarchitecture*. For instance, the x86 ISA implementation is different in Intel, AMD, and Centaur microarchitectures. The processor contains the circuit logic to execute the set of instructions. ASICs have unique ISAs usually not shared publicly.

Two general types of instruction sets are the complex instruction set computer (CISC) and the reduced instruction set computer (RISC). The CISC ISA aims to execute multiple low-level operations per instruction. The RISC ISA is smaller and simpler than the CISC ISA and aims to provide higher IPC rates. The most common instruction sets and the typical devices that use them are as follows:

• CISC x86 ISA in computer processors from laptops to supercomputers;

- RISC Arm ISA in smartphones with some adoption in laptops and single-board computers and starting to enter the server market by Ampere, AWS-Graviton, Marvell, and Huawei;

- RISC open-sourced RISC-V ISA in academia with some small traction in production at Alibaba and SiFive; and

- RISC Power ISA in IBM POWER microprocessors and some supercomputers.

There are different ways to parallelize a kernel in hardware, such as with SIMD/SIMT instructions, multicores, or systolic architectures. Also, model parallelism techniques, discussed in Section 6.3, can distribute the kernel's computations among multiple nodes.

Single instruction, multiple data (SIMD), and single instruction, multiple threads (SIMT) (coined by Nvidia), are used by CPU and GPU vector processors, respectively. In CPUs, a SIMD instruction is concurrently applied to all the values in the respective registers within an execution unit (EU) in a core. To illustrate, an AVX-512 instruction execution unit (EU) may take two 512-bit input registers, each with 16 $fp32$ values, and computes the element-wise product across the registers and stores the resulting 16 $fp32$ values in another 512-bit register. GPUs generalize SIMD with SIMT; rather than apply an instruction to data in registers, GPUs apply an instruction across multiple threads (a *warp* or 32 threads in Nvidia GPUs and a *wavefront* or 64 threads in AMD GPUs). Specifically, GPUs use coalesced loads, where different values in the same cache line are concurrently accessed and used by the threads in a warp or wavefront.

SSE, MMX, AVX, AVX-2, and AVX-512 (sometimes called AVX-3) are SIMD instruction extensions to the x86 ISA, and NEON and the Scalable Vector Extensions (SVE) are SIMD instruction extensions to the Arm ISA (do not worry if you are unfamiliar with these ISAs). The primary differences between these ISA extensions are the number of supported instructions and the data size that the instruction can be concurrently applied. For example, AVX-512 has more instructions than AVX-2 and concurrently operates on 512 bits, whereas AVX-2 operates on 256 bits.

Nvidia provides a pseudo-assembly language virtual ISA called the *Parallel Thread Execution* (PTX). Compilers, such as GCC (detailed in Section 8.2), generate PTX code. PTX code requires using Nvidia's NVCC compiler to access the physical ISA known as SASS to generate an executable binary [Nvi15]. Recent AMD GPUs use the Vega ISA, RDNA ISA, or CDNA ISA.

Simultaneous multithreading (SMT), called hyper-threading for Intel processors, is used in CPUs to run two (and potentially four or more) threads in one core to utilize better the EUs that may otherwise be idled. For well-optimized kernels, however, an EU may not sit idle, and using two threads may not provide a significant gain in performance. In high OI kernels, enabling SMT could reduce the performance due to the thread switching overhead. Experimentation is required to assess the gains or losses of SMT on a particular workload.

Another set of instructions designed to exploit instruction-level parallelism is the very long instruction word (VLIW) instructions, where multiple instructions execute in parallel. VLIW processors work best with regular, predictable code for the compiler to extract the required level of parallelism. The retired Itanium, and today's Habana AI processors as well as Google's TPU v2 (and perhaps v3 and v4) use VLIW SIMD vector processors.

Dataflow parallelism uses systolic architectures (also called dataflow architectures or dataflow processors) with multiple simple processing engines (PEs). A PE performs a simple computation, such as a MAC, and passes the result to its neighbor PE. The collected work across all PEs results in high throughput. Given the simple circuitry design, dataflow architectures can be power-efficient. In a systolic array, the PEs connect in a mesh pattern; the shorter wires connect nearby PEs and provide high bandwidth at much lower power than longer wires. Dataflow parallelism is adopted in specialized hardware discussed below, including in domain-specific circuitry added to CPUs and GPUs, such as Intel's AMX and Nvidia's tensor cores. Dataflow processors work best with regular, predictable code. Using systolic architectures (and SIMD and SIMT) near peak performance, requires a mature compiler or program that considers the memory hierarchy. A minor mismatch from memory access to the systolic dataflow processor can lead to orders of magnitude of slower performance.

A **central processing unit (CPU)** consists of RAM, registers, and execution units. RAM holds both the program instructions and the data. A server CPU typically has faster but fewer cores compared to a GPU or a dedicated DL accelerator. It may better balance complex workloads: the parallelizable code can benefit from the many CPU cores, and the serial code can benefit from the single-core high-frequency performance. Note that the execution time does not decrease linearly with increased core count, per Amdahl's law. A CPU provides maximum flexibility and is typically simpler to program than other hardware. It has built-in logic to exploit control-flow, including branch prediction. This flexibility comes at the expense of higher power consumption to decode and execute the instructions in each core. Embarrassingly parallel workloads with static graphs do not require many of the capabilities of the CPU, and a dedicated processor should provide higher performance per watt.

A **graphical processing unit (GPU)** consists of RAM, registers, and compute units. GPUs are designed for embarrassingly parallel tasks, initially targeting image manipulation by simultaneously applying an operator to each pixel or group of pixels, and later targeting DL matrix multiplications and convolutions. A difference between a CPU and a GPU core is that the CPU core can decode and execute an instruction independently of the other core. A GPU core executes the same instruction as the other cores in their group, known as a warp and wavefront by Nvidia and AMD, respectively. The CPU cores provide more flexibility than GPU cores, and the GPU cores provide higher energy efficiency than CPU cores.

A CPU core is an independent processor with dedicated ALUs, control logic, local SRAM with a dedicated $L1$ cache, and multiple registers shared only between the SMT threads (when SMT is enabled). A GPU core cannot operate independently of other cores; it has dedicated

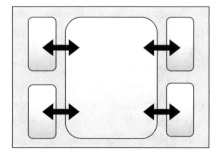 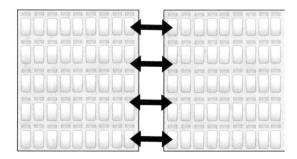

Figure 7.9: The memory designs under the same power consumption range from using (left) HBM and small local SRAM to using (right) multiple SRAM units that take most of the silicon area and no DRAM. Blue rectangles represent the memory and yellow rectangles the compute units.

registers but not dedicated SRAM; instead, it shares the memory with all the cores in the warp or wavefront. Given the limitations of a GPU core, some literature refers to them as threads. The warp or wavefront can be thought of as a core with massive SMT capabilities. Compared to CPUs, GPUs use much larger register files similar in sizes to a CPU's LLC to support the massive SMTs at higher throughput at the expense of higher latency.

A typical bottleneck is the limited memory bandwidth. Increasing the SRAM associated with every compute unit or PE can mitigate this. Design choices range from Nvidia's V100 with large HBM2 and small local SRAM to Graphcore's Colossus with no DRAM and large SRAM units that take most of the silicon area, as illustrated in Figure 7.9, and to emerging in-memory processing technology. The design choices affects the batch size required to achieve high efficiency. Hardware with more local SRAM can have higher compute utilization with small batch sizes, which can benefit both training and inference. Training with small batch sizes requires less hyperparameter tuning to converge. Inference with small batch sizes (often a batch size of one) is typical to meet latency constraints.

A **field-programmable gate array (FPGA)** is a type of hardware with some small compute elements (logic blocks), such as memory, registers, lookup tables, and macro functions, and whose connectivity is reconfigurable and can be programmed. This programmability is beneficial to adapt to new workloads that require different hardware characteristics. Also, FPGAs are used to simulate ASICs and other processor designs before building them. Two challenges with FPGAs are the long compilation time (several minutes to hours) to reprogram the logic gates and the limited DL software tools.

A **coarse-grained reconfigurable array (CGRA)** is also a type of programmable hardware [LZL+19]. A CGRA can be thought of as an FPGA with coarser reconfigurability. Thus, in theory, a CGRA provides easier programmability but less flexibility compared to an FPGA. In practice, CGRAs have limited adoption due to the limited software tools.

A **digital signal processor (DSP)** is a specialized, low-latency microprocessor with a specialized ISA optimized for frequently used functions in signal processing, like convolution. Modern DSPs are modular in that they may have a base ISA that is consistent, and an extension ISA that is specific to the type of processing (for instance, for image, audio, and network signals). Unlike a CGRA, a DSP is not reconfigurable. DSPs are programmable but require a good compiler for high performance. DSPs are typically used in combination with other hardware in a heterogeneous design.

An **application-specific integrated circuit (ASIC)** provides the best performance for a specific application but is the least flexible. ASICs have limited control logic and depend on the programmer or compiler to manage data movement. Achieving high-utilization requires experienced low-level programmers or a matured DL compiler. Current DL compilers are still immature and require significant time to map the kernels to execute efficiently in hardware. Given the software complexity, ASICs work best with regular, predictable code. Some newer models have dynamic graphs with complex datapaths that are difficult to compile efficiently. ASICs are often used as part of a DL design with other architectures to handle the computationally intensive operators.

Most ASICs use dataflow architectures for MAC computations. A recommended high-level design for an ASIC is to pack as many transistors as possible into a die for MACs operators (the die size and power constrained by the deployment environment) to support matrix-wise operators. Then, use some of the silicon for element-wise operators, matrix transposes, and I/O, and use most of the rest for SRAM. The processor should operate at or slightly above the $V_{\min}$ voltage to ensure the highest ops/s per watt. Increasing the frequency past $f_{\min}$ increases the power with the cube of the increased frequency (see Section 7.1).

There are various ways to implement MACs with dataflow parallelism. Chen et al. and Sze et al. provide a detailed review of various dataflow architectures [CES17, SCY+20]. These architectures have an array of PEs connected via a network-on-chip (NoC) with global SRAM memory, as illustrated in Figure 7.10 for a 3×3 array (in practice, the arrays are larger). The PE array gets the activations (Act), weights, and accumulated sum from the global SRAM. Each PE contains the ALU or FPU logic to perform MAC operations, a local control unit, and may have a local SRAM scratchpad (Spad). The MAC unit multiplies a set of weights and activation and adds the result to the accumulated partial sum.

There are four types of dataflow architectures: no local reuse, weight-stationary, output-stationary, and row-stationary [CES17].

No Local Reuse maximizes the size of the global SRAM by not having local PE memory. The weights and activations pass from the global SRAM to each PE, with passes the accumulated sum to its neighbor along a row of PEs, as illustrated in Figure 7.11.

Weight-Stationary maximizes weight reuse by storing the weights in the PE's local memory. An activation is broadcasted to the relevant PEs, and the accumulated sum flows from each PE to its neighbor along a row of PEs, as illustrated in Figure 7.12. This data flow works well

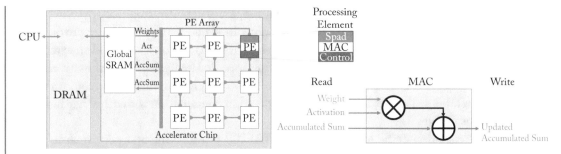

Figure 7.10: An accelerator chip with a 3×3 array of PEs. Each PE has a MAC unit that multiplies a set of weights and activations and adds the result to the accumulated sum. Based on [SCY+17].

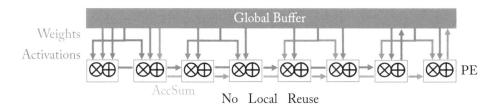

Figure 7.11: No local reuse dataflow architecture. Based on [SCY+17].

for traditional convolutional layers that reuse the weights. It is not efficient for fully-connected layers or convolutional layers with limited weight reuse, such as 1×1 convolution or depthwise separable convolutions.

Output-Stationary maximizes reuse of the accumulated sums by storing them in the PE's local memory. A weight is broadcasted to all the relevant PEs. The activations flow from each PE to its neighbor along a row of PEs, as illustrated in Figure 7.13.

Row-Stationary maximizes reuse across weights and activations. The accumulated sums flow from the bottom to the top columns, as illustrated in Figure 7.14. Row-Stationary, proposed by Chen et al., provides the best performance per watt for convolutions and fully-connected layers [CES17, CES16].

An operation may not distribute evenly across all the PEs in the array. dMazeRunner efficiently explores the various ways to split computational kernels in a dataflow accelerator [DKA+19].

ASICs can also be customized to better support sparse matrix multiplications. Nvidia researchers demonstrated the benefits of sparse multiplications with the ExTensor accelerator that rapidly finds intersections of nonzero operands and avoids multiplies by zero [HAP+19].

Compute-in-memory and neuromorphic processors are two different designs; both have challenges and no adoption in production. A **compute-in-memory processor** uses analog com-

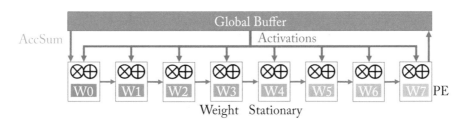

Figure 7.12: Weight-stationary dataflow architecture. Based on [SCY+17].

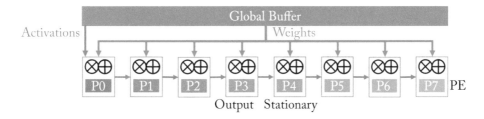

Figure 7.13: Output-stationary dataflow architecture. Based on [SCY+17].

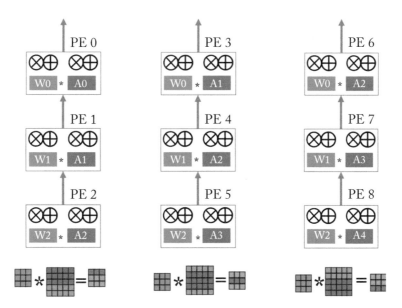

Figure 7.14: Row-stationary dataflow architecture. Based on [SCY+17].

putations [TAN+18]. The tunable resistors represent the weights, the voltage represents the activations, and the measured output current represents the accumulated sum. While very efficient to multiply values, this type of hardware requires meticulous weight tuning to mitigate statistical performance degradation. Furthermore, the expensive digital-to-analog (DAC) and analog-to-digital converter (ADC) limit overall power savings.

A **neuromorphic processor** is a type of brain-inspired processor that attempts to reduce power consumption in comparison to other DL hardware [RJP19]. It uses spiking neural networks (SNNs) at very low power. However, its success is limited to simple domains and has lower statistical performance than traditional artificial neural networks (ANNs). The input-to-output function is nondifferentiable, preventing the use of the backpropagation algorithm, except as an approximation [LLH+19, ZL19].

7.5   HIGH-PERFORMANCE INTERCONNECTS

Training and inference benefit from high-performance interconnects for data ingestion, model parallelism, and data parallelism. The types of interconnects discussed in this section are host-to-device, such as PCIe and CXL and device-to-device and host-to-host, such as InfiniBand, OmniPath, and Ethernet/IP. Hyperscalers typically use commodity interconnects, such as Ethernet, to lower operational costs, unlike in HPCs, where it is common to use custom fabrics [ALV08].

Host–device interactions focus on efficiently allocating the computations between the host and the device. For instance, the host may be responsible for data cleaning and preprocessing before handling the data off to the device for dense computation. Host-to-host and device-to-device interactions focus on supporting parallel computation involving multiple instances of the host or the device, such as for distributed training. Host–storage interactions focus on feeding the training dataset to the host or hosts.

Serializer/Deserializer (SerDes) is used for data transfers. A SerDes chip converts parallel data to serial data to be transmitted over a much higher speed connection and vice versa. Multiple SerDes interfaces are often used in a single package. SerDes standards include the Peripheral Component Interconnect Express (PCIe) bus and Ethernet. Proprietary SerDes are used in Nvidia's NVLink, Nvidia Mellanox InfiniBand, and AMD's Infinity Fabric.

The PCIe bus is an industry bus standard that connects various devices, such as the host CPU, GPUs, accelerators, Ethernet, and other I/O devices commonly found in modern computers, such as SATA and USB. PCIe interfaces can be configured in different number of lanes (or widths) based on the intended application of the computer system design. The theoretical peak bandwidths per direction using 16 lanes (written as ×16) are doubling in almost every generation as follows:

- PCIe 3.0 ×16: 16 GB/s (most common)

- PCIe 4.0 ×16: 31.5 GB/s (recently available)

- PCIe 5.0 ×16: 63 GB/s (future)

The Compute Express Link (CXL) is a new high-speed CPU-to-device interconnect that can maintain memory coherency between the CPU and device memory. The technology is built on PCIe and leverages the PCIe 5.0 physical and electrical interface, meaning CXL and PCIe are electrically identical. They both use the same PCIe form factor CEM, which allows using a single slot in a server for PCI or CXL devices. However, CXL and PCIe have different and incompatible protocols.

High-performance network protocols which support efficient scaling of accelerators include RDMA over Converged Ethernet (RoCE), InfiniBand, iWARP, and Omni-Path. These network protocols allow remote direct memory access (RDMA). RDMA utilizes the memory subsystem and maps a remote memory region to be accessible directly by the network interface controller (NIC) to obtain higher bandwidths without being bottlenecked by connection to the host, or via a software construct to shared memory that consumes CPU cycles. It is up to other technologies to allow this mapping and get the RDMA to the correct destination. In the case of RDMA to CPU memory, it is through the memory controller in the CPU. In the case of two devices that support PCIe peer-to-peer (P2P), it is a P2P transaction between the devices. NVLink is a different interface and protocol which enables direct GPU to GPU memory sharing and communication between Nvidia GPUs. AMD GPUs use Infinity Fabric. Other accelerators are adding interfaces, often proprietary and optimized for their specific architecture, to allow for direct device-to-device memory sharing or communication to provide high bandwidth, multidevice scaling.

There is a difference between the protocol for the chip-to-chip interconnect, such as PCIe and Ethernet, and the actual physical interconnect (PHY), which can carry multiple types of protocols on top of it. The PHY Interface for PCIe (PIPE) is used for PCIe, CXL, and USB 3.0 SuperSpeed. The PCI-SIG and the IEEE 802.3 define and manage the PCIe and Ethernet standards, respectively. The Optical Internetworking Forum (OIF) promotes the development of interoperable SerDes devices.

7.5.1   PHYSICAL NETWORK TOPOLOGIES

There are different device-to-device and host-to-host physical network connection, as shown in Figure 7.15. In the early days of high-performance computing (HPC) interconnects, low-radix topologies were the norm. Higher radix network topologies are becoming more common as the pin bandwidth increases and can be efficiently partitioned across multiple ports [KDS+08, KDT+05].

Distributing the training of large models across several nodes requires communicating a large number of weights or activations. Therefore, high node-to-node bandwidth is beneficial. The optional choice of the communication primitive algorithm is dependent on the physical network topology. High-radix topologies are preferable when the bandwidth can be effectively utilized, as is the case in the AllReduce-Ring algorithm, but not in the other AllReduce algorithms analyzed in Section 5.4.

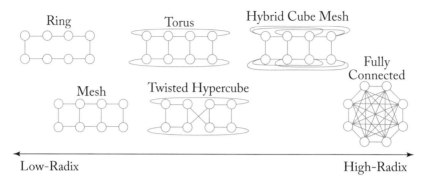

Figure 7.15: Examples of topology designs using 8 nodes. High-radix topologies provide lower communication latency across nodes. Based on [NKM+20].

7.6   PROCESSORS IN PRODUCTION

There are several processors in production and development for DL, and many more in research. The number of processors is likely to expand in the next few years and later consolidate, similar to what happened with the DL frameworks, as detailed in Chapter 9. In this section, we discuss some of the most prominent processors in production and development.

CPUs are widely used for inference. Server CPUs are also used to train small models, models requiring large memory capacity, and large models during off-peak hours [Haz18]. CPU designers continue to add more cores; the 2nd-generation Intel Xeon Scalable processors has up to 56 cores and the AMD EPYC 2nd gen has up to 64 cores [Int19, Amd19]. In addition, CPUs are incorporating specialized circuits and instructions to accelerate DL computations and reduce power overhead. Apple's A13 Arm-based CPU (used on the iPhone 11) has two cores with AMX blocks dedicated for matrix multiplications [Wik19]. Arm introduced additional instruction sets to accelerate $bf16$ multiplications on the Armv8-A architectures [Ste19b]. Intel introduced extensions to the AVX512 ISA to support $int8$ and $bf16$, and is adding the Intel Advanced Matrix Extensions (Intel AMX) (not related to Apple's AMX) with 2D registers and dedicated $bfloat16$ and $int8$ acceleration starting with the Intel processor codenamed Sapphire Rapids [Int19, Int20, Int20b]. IBM POWER10 cores have an embedded Matrix Math Accelerator to accelerate $fp32$, $bf16$, and $int8$ operations [Ibm20].

GPUs have thousands of threads designed for embarrassingly parallel tasks, such as large batch GEMMs and convolutions. Some GPUs are also incorporating specialized features. Starting with the Volta microarchitecture, Nvidia added tensor cores to increase $fp16$ MACs throughput. The Turing microarchitecture added $int8$ and $int4$, and the Ampere microarchitecture added $bfloat16$ and $fp19$ support to the tensor cores. Nvidia also designed an RISC-V-based dataflow architecture RC-18 for fast low power inference possibly available in the (tentatively called) Hopper microarchitecture [VSZ+19, Mor19].

AMD also produces GPUs but has not found the same success as Nvidia. The primary gap AMD GPUs have is the limited DL software ecosystem. AMD shifted from a single family of microarchitectures, the graphics core next (GCN), which expanded five generations, to two families: the RDNA and CDNA microarchitectures. The Vega GPUs use the GCN 5th generation. The Navi GPUs use the RDNA graphics microarchitecture. The Arcturus GPUs use the CDNA compute microarchitecture starting with the Radeon Instinct MI100. AMD added *bfloat16* support to their ROCm MIOpen DL software library, a likely indicator of support in the CDNA architecture.

Intel plans to release a discrete general-purpose GPU for HPC modeling and DL training based on Intel's X^e architecture, codenamed Ponte Vecchio. Among the adopters is the Aurora supercomputer at Argonne National Laboratory [Kod19, Int19b].

FPGAs have some adoption in production for ML workloads. For instance, Microsoft Brainwave and Azure provides ML services on FPGAs. The primary FPGAs makers are Intel (after acquiring Altera) and Xilinx. Both offer high-level software libraries to facilitate using FPGAs with a predefined set of DL primitives: OpenVINO and Xilinx ML Suite.

Xilinx introduced the Adaptive Compute Acceleration Platform (ACAP) combining an Arm-based CPU, a vector processor engine, and an FPGA in a high-bandwidth NoC [Xil19]. Intel has demonstrated the benefits of enhancing FPGA with specialized tensor functions [NKJ+19].

Established silicon vendors, startups, and hyperscalers are developing a spectrum of specialized DL processors, often using a mixture of ASICs and DSPs. Google was the first to succeed at scale with the Tensor Processing Unit (TPU). The TPU is used for a range of DL workloads by several companies, including Google, eBay, and the Mayo Clinic [ZZZ+18, Hou19]. TPU v1 is designed for *int8* inference workloads, and v2–v4 are designed for training, although inference is also supported at lower latency and higher throughput than v1 [JYK+20]. TPU v2–v4 adds *bf*16 support, HBM, and improves the scalar, vector, and matrix (systolic array) units. TPU v2 (and possibly other generations) uses VLIW and software-managed instructions. Other specialized processors developed by hyperscalers for their cloud services are: AWS Inferentia, Alibaba Hanguang 800 NPU, and Baidu Kunlun [Ama19, Ama19b, ZLH+19, JHJ+20, Bai20].

Specialized processors developed by established and merchant market silicon vendors include the Intel Habana processors, the Intel Movidius Myriad-X edge inference processor, the Qualcomm Cloud AI 100 inference processor, and the Huawei Ascend 910 training processor. The Habana Goya inference processor and the Habana Gaudi training processor both use a scalable architecture of C-programmable VLIW SIMD vector processors called Tensor Processing Cores (TPC).

Prominent startups with unique designs are Cerebras System, Graphcore, SambaNova, and Groq. Cerebras Systems CS-1 is designed to facilitate distributed training. Rather than manufacturing multiple chips from a silicon wafer and then using SerDes to connect them, CS-1 uses one gigantic TSMC 16 nm chip, called a *wafer scaled engine* (WSE), which uses an entire silicon wafer. Thus, CS-1 consolidates multiple chips into a single chip with $1.2T$ transistors. The design has redundancy to account for defects in parts of the wafer, typical in the manufacturing process. The Gen-2 WSE, introduced in Aug. 2020, more than doubles the Gen-1 cores and transistors.

Graphcore released the TSMC 7 nm MK2 2nd-generation Intelligence Processing Unit (IPU) in July 2020. The majority of the die is dedicated to the 900 MB SRAM interlaced throughout the computing clusters totaling $fp16 \rightarrow fp16$ 250 TOPS. The intent is that the SRAM is large enough to hold the entire model on a chip facilitating high-efficiency small batch training and inference. Each PCIe card contains two IPUs with high chip-to-chip bandwidth so that a model can be split between them. Microsoft has deployed the 1st-generation IPU processors in the Azure cloud [Gra19]. It is unclear if the $fp16 \rightarrow fp16$ MACs (rather than $fp16 \rightarrow fp32$ MACs) may present a convergence challenge for some workloads, or if this increases the complexities to port existing code from a different architecture into MK2.

SambaNova is developing the Reconfigurable Dataflow Unit (RDU) designed as a dataflow architecture, possibly based on the Plasticine CGRA architecture [PZK+17]. SambaNova developed the DataScale System with petaflops of performance and terabytes of memory capable of efficiently training a 100 billion parameter model.

Groq's Tensor Streaming Processor (TSP) supports $fp16$ and $int8$ multipliers. It offers 820 $int8$ TOPS and 220 MB SRAM. Similar to Graphcore's MK2 IPU, the large SRAM capacity is beneficial for small-batch processing. The TSP can be reconfigured at every cycle by software using ahead-of-time compilation [ARS+20]. It is deployed at Nimbix cloud [Nim20].

CS-1, Graphcore's IPU, and Groq's TSP designed chips with large SRAM to reduce the cost and time to fetch data from main memory. Training, however, requires storing the activations, which can exceed the available SRAM. To fully realize their potential, pipeline parallelism with gradient checkpoints, introduced in Section 5.2.1, is likely required to avoid main memory accesses. While pipeline parallelism suffers from similar drawbacks as Async SDG, recent algorithmic improvements are showing promising results on small-scale experiments [KCV+20]. Further advancements in pipeline parallelism are necessary to deliver on these chips' potential.

7.7   PLATFORMS STRENGTHS AND CHALLENGES

DL platforms must balance memory, bandwidth, general-purpose compute, and dedicated compute to effectively train and serve DL workloads with heterogeneous characteristics. State-of-the-art recommender models and emerging GNN models require more general-purpose compute than NLP or computer vision workloads. A platform should also be programmable and sufficiently flexible to adapt to novel workloads with different demands. Given the larger complexities of training, we focus on training platforms in this section and review the Nvidia DGX, Google TPU POD, Habana HLS-1, Graphcore M2000, Cerebras CS-1, AMD CDNA-based, Hyve Solutions Catalina, and Facebook Zion platform.

Nvidia markets GPU-accelerated servers (DGX) and racks (POD). NVLink with NVSwitch connects the GPUs within a server, and InfiniBand connects racks of servers for scale-out. The intra-server NVLink connectors facilitate model parallelism as the entire server can be treated as one large compute unit while the inter-server InfiniBand primarily facilitates data parallelism. The SuperPOD-based Selene supercomputer ranked 7th in the TOP500 June 2020 list. The platform's biggest strength and enormous competitive advantage is the mature software, and the robust ecosystem including multiple active online forums. Also, the extensive market adoption of Nvidia GPUs results in better economies of scale lowering production costs. The Ampere microarchitecture offers multi-instance GPU (MIG); a single A100 can be partitioned into up to seven virtual GPUs, each of which gets its dedicated allocation of cores, L2 cache, and memory controllers. This partition is particularly beneficial for small-batch inference, as using the entire GPU would result in low compute utilization.

Google designed the TPU POD connecting many TPUs with custom interconnects (the chips have integrated routers) to work on a single workload offering over 100 PFLOPS on the TPUv3 POD. The TPUv2 POD uses a flat 2D torus physical network topology across all 16×16 TPUs. The TPUv3 POD has 1024 TPUs, and possibly uses a 32×32 2D torus network topology. Google demonstrated the fastest training time in six of the eight MLPerf workloads (the DGX A100-based POD won the remainder two) in the July 2020 submission using a TPUv4 POD. The MLPerf benchmark is discussed in Section 7.8. The flat network topology does not differentiate between local and global bandwidth since the fabric is organized as a single, global topology. The biggest competitive advantage is the first-to-market data center DL accelerator as an alternative to Nvidia GPUs. Also, having a dedicated DL processor with software-managed memory enables Google to design the TPU with comparable performance to a contemporary Nvidia GPU but using an older semiconductor process, which lowers costs. Facebook and Google are collaborating to add TPU support to PyTorch. PyTorch and Tensor-Flow are the dominant DL libraries in industry and are detailed in Sections 9.2 and 9.3. The primary inference challenge with the TPU design is the large matrix multiply units (256×256 or 128×128), which are beneficial for large-batches but typically result in low-utilization for small-batch processing. While Google can aggregate hundreds of inference requests in real-time into a batch, non-Google TPU users with much smaller scales may not be able to carry out large-batch inference.

Habana, an Intel company, developed the Habana Labs System-1 (HLS-1) with eight Habana Gaudi cards. Gaudi integrates RDMA engines on the chip and uses industry-standard Ethernet as the physical interface without requiring a NIC to communicate within and across platforms. Each Gaudi in HLS-1 uses ten 100 Gbps RoCE channels: seven for intra-box connectivity and three for scale-out [Hab19b]. A strength of HLS-1 is using Ethernet, which facilitates using the existing Ethernet infrastructure. Also, the HLS-1 platform does not include a CPU; it provides PCIe switches for a host interface, and the user determines the right CPU to Gaudi ratio. Thus, the HLS-1 platform decouples the management traffic to PCIe and the scale-out traffic to Ethernet, which mitigates PCIe and scale-out bottlenecks. Lastly, the Habana training and inference processors share a similar design and programming model, which reduces the training to inference switching costs.

Graphcore developed the IPU-Machine M2000 platform. Each platform has 4 MK2 IPU devices disaggregated from the CPU (similar to HLS-1 in this regard). Graphcore developed the reference architecture IPU-POD64 with 16 M2000 platforms totaling 64 MK2 IPUs. Graphcore offers a hybrid model with proprietary IPU-Links connecting the IPUs within an IPU-POD64 in a 2D torus physical topology and provides IPU access over RoCE to scale outside an IPU-POD [Gra20]. A strength is the large IPU SRAM capacity for high-efficiency, small-batch processing. Also, Graphcore is collaborating with Microsoft, a large company with a diverse set of workloads and large datasets, which can help Graphcore mature their software on relevant workloads. One challenge is the lack of *fp*32 accumulation in the MK2 IPU *fp*16 MACs, which may impact training accuracy in some workloads.

Cerebras System markets a clusters of CS-1s. This platform is deployed at Argonne National Laboratory (ANL) [Fel19b], and two systems at the AI supercomputer Neocortex in the Pittsburgh Supercomputing Center (PSC). The strength of CS-1 is consolidating multiple DL chips into a gigantic processor, which provides high-bandwidth for distributed training and large SRAM capacity. Similar to the MK2 IPU, the large SRAM facilitates small-batch processing. The primary challenges are the inability to modify the dedicated compute, bandwidth, and CPU ratios, and effectively utilizing all the available compute.

AMD GPU-accelerated servers use the AMD Infinity Fabric Link to directly connect up to 2 GPU hives of 4 GPUs per hive. AMD CDNA-based GPUs are accelerating the El Capitan supercomputer in 2023, and use the high-bandwidth, low latency 3rd-generation AMD Infinity Architecture links between 4 GPUs and 1 CPU. AMD is growing its ecosystem with hyperscalers, which may facilitate entry into the DL market, and having a system with a CPU and a GPU developed by the same company may simplify running workloads on the heterogeneous system.

Hyve Solutions designed the Catalina platform using eight sockets of 3rd-generation Intel Xeon Scalable processors, which double and quadruple the *bf*16 and *int8* theoretical peak compute, respectively, over AVX512 *fp*32. The strength is fungible computing. It can be used for DL training and inference, traditional ML, HPC, and general-purpose workloads. The challenge is the lower performance compared to a dedicated DL training platform for users who can maintain the dedicated platform utilized the majority of the time.

Facebook designed the large-memory unified training platform Zion that consists of both CPUs and accelerators and provides disaggregated memory, compute, and network components, allowing each to scale independently [NKM+20]. The CPU chassis Angels Landing is composed of four dual-socket CPU modules. The accelerator chassis Emeralds Pools uses up to eight vendor-agnostic OCP accelerator modules (OAMs) with high-speed high-bandwidth interconnects. The OAM form factor, co-developed by Facebook, Baidu, and Microsoft, allows hardware vendors to develop accelerators with a standard specification. The platform has three types of interconnect fabrics—the CPU fabric, the accelerator fabric, and the PCIe interconnect that provides connectivity between CPUs and accelerators. The strength of Zion is its capacity to support a broad spectrum of DL workloads. Those with significant memory demands can use the large DRAM capacity in the eight-socket platform, and those with significant compute and memory bandwidth demands can benefit from the HBM and specialized compute in the accelerators. The challenge is the software complexity to move the data and use all the available compute efficiently. All platforms discussed in this section share this challenge, and it becomes more prevalent the more heterogeneous is the platform.

7.8   EVALUATING DEVICES AND PLATFORMS

An important metric is the total cost of operating the hardware, including the hardware's cost, the maintenance over its lifetime, and the software engineers to program the hardware. The ease-of-programming to support a wide spectrum of topologies at high performance is an important part of the evaluation, which cannot be overstated.

Unsurprisingly, different products have different use cases. Training hardware is optimized for throughput. Inference hardware is optimized for latency, and edge hardware is optimized for power and size. Also, a topology may have an affinity to a specific processor. While the number of weights, number of layers, and size of activations can affect performance, it is essential to evaluate a topology in the context of a particular hardware architecture. Critical metrics for a topology and hardware pair are statistical performance, computational performance, and power consumption.

The following platforms facilitate evaluating DL hardware. FireSim is an FPGA-based hardware simulator available on AWS FPGA instances; the Nvidia DL Accelerator (NVDLA) is integrated on FireSim [KBA18, FHY19]. The SMAUG and Eyexam packages model the performance of a topology on an accelerator design [XYB+19, CYE+19]. The ParaDnn tool is used to benchmark DL platforms against TPUs, GPUs, and CPUs [WWB19]. Wang et al. and Dai et al. provide performance comparisons on various hardware targets [WWS+19, DB19].

The community is developing a suite of benchmarks, such as DeepBench to evaluate primitives, DAWNBench to evaluate performance and cost on a public cloud service, and MLPerf [Dee19, Daw20, Mlp18]. MLPerf is the most popular benchmark backed by a consortium made up of some of the biggest companies in DL and evaluates performance across well-established models.

While benchmarks are essential to evaluate DL systems, a potential danger is overfitting the hardware and software designs to a benchmark. Most benchmarks focus on the past rather than the future. It may be wise to develop benchmark metrics that provide a measure of the platform's programmability and flexibility to support a diverse range of workloads. This metric should include the compilation time required to obtain high efficiency on these models.

In this chapter, we reviewed the basics component of DL hardware and the ample design space for training and inference. We detailed why a smaller numerical representation requires less silicon and power and discussed some of the performance vs. ease-of-programmability trade-offs across various hardware designs used for DL: CPUs, GPUs, DSPs, FPGAs, CGRAs, and ASICs, as well as heterogeneous designs. We recommended a high-level ASIC design to maximize OPS per watt with sufficient die area for SRAM and other critical circuitry. We discussed the high cost of accessing DRAM memory and the pipeline parallelism related challenge that large SRAM training processors have to overcome to deliver on their potential. We highlighted prominent DL processors and platforms in production and development and emphasized the need for a flexible and programmable platform that supports a broad spectrum of workloads to gain wide adoption. Given the approximately two to three years to bring hardware from concept into production, a platform needs to account for unforeseen algorithmic and model innovations. A flexible platform design may include disaggregated CPU to accelerator ratio, a standard form factor module, and an industry-standard interconnect to scale out the architecture. This flexibility facilitates the evaluation and adoption of heterogeneous processors, which is important to data center managers to avoid being locked into one vendor. Moreover, while flexibility comes at the expense of some performance, given the rapid algorithmic innovation, the benefit is likely worth this price. We also discussed the challenges with software-managed memory and the complexities to extract high performance; the program needs to efficiently map to the target hardware via compilers accurately matching memory accesses to the ALU or FPU logic inputs and results. In the next chapter, we review the basics of compilers and describe the standard compiler optimizations passes for DL workloads.

CHAPTER 8

Compiler Optimizations

At the core of the software stack are compilers to transform the programmer's high-level code into executable code that runs efficiently on a target device. Programmers use a variety of languages to code at various levels of abstraction. A programming language is a formal language used to write code, such as for functions and algorithms. High-level languages are independent of a hardware target and include C, C++, Python, Java, Javascript, CUDA C/C++, Swift, and Julia. Assembly (asm) is a low-level language that targets a specific instruction set architecture (ISA). In between are intermediate languages that are assembly-like in format but general enough for execution on different ISA, such as LLVM IR, various Multi-Level IR (MLIR) dialects, and PTX for Nvidia GPUs.

Programming languages have a set of specifications or rules that dictate what the outputs should be for a given input. The output also depends on the dynamic conditions of the running program. The approaches to implement a programming language are interpretation, compilation, or a mixture of both. The terms *interpreted language* and *compiled language* denote that the default or canonical implementation of that language uses an interpreter or a compiler, respectively. For some languages, the canonical implementation is the only implementation, while others like Python have multiple implementations (more on this below).

An interpreter is a computer program that directly executes the code for a particular language. That is, the code does not map to machine code. The processor executes (runs) the interpreter, and the interpreter reads and generates the output for the interpreted language according to the interpreted language's specifications and rules. The interpreter's source code (the program that is executed) can be a different language than the interpreted language.

A compiler is a computer program that transforms code between two languages or within a language. The compiler runs various optimization passes to improve the execution time and simplify the code. Alternatively, the compiler may only focus on code canonicalization, which transforms the code into more rigid patterns removing unnecessary variations. The compiled code is passed to an interpreter or directly to the processor when it is machine code (in this case, the processor can be thought of as the interpreter of the machine code).

Often, before an interpreter executes a high-level code, the code is first dynamically (just-in-time) compiled into *bytecode*, which is a compact language or efficient intermediate representation. This compilation is usually a minor transformation to make it easier for the interpreter to parse the code. Typically, more compilation (optimization passes) leads to faster execution; however, this comes at the expense of longer build time.

Let us look at the Python language as an example of a language with various implementations, and focus on two: CPython and PyPy. CPython is an interpreter implementation and the canonical (default) Python implementation. Python programmers that have never heard of CPython likely use the CPython interpreter. Like other interpreted languages, the Python source code or Python command, when used interactively by the programmer, is transformed into bytecode. Then, this bytecode is interpreted by CPython one command at a time. PyPy is an interpreter and a JIT compiler (more on JIT below) Python implementation.

Compilers *lower* (this is compiler parlance for *transform*) code from a higher-level language to a lower-level language, for instance, from C++ to x86 machine code. Compilation to machine code that happens before runtime (execution) is known as Ahead-of-Time (AOT) compilation. Compilation to machine code that happens during runtime is known as Just-in-Time (JIT) compilation. AOT improves the performance for static graphs at the expense of longer compile times.

A JIT compiler is a computer program that compiles to machine code at runtime. Using a JIT compiler can significantly increase startup time. To mitigate, JIT compilers are typically used alongside an interpreter for runtime profile-guided optimizations, also known as adaptive optimizations. As the interpreter executes the source code (or, more precisely, the bytecode), the interpreter tracks repetitively used sections and triggers the JIT compilation for these sections into higher-performing machine code. The compiled code is cached, and the interpreter can then alternate between the usual execution of bytecode and the execution of the JIT code.

An intermediate representation (IR) is a data structure or graph representing the required operations for a particular program. Compilation may use several levels of IR, progressively lowering on each pass. A high-level, hardware-independent IR may contain control-flow tokens, such as `for`, `if`, and `while`. A low-level, hardware-independent IR may look similar to assembly language while still being generic enough not to be tied to a specific hardware implementation to simplify the next stage of compilation. Bytecode is an example of an IR.

Two common properties of some IRs are static single-assignment (SSA) form and three-address code (TAC). SSA requires that each variable (called a typed *register*) is assigned precisely once (it is not mutable), and every variable is defined before it is used, which facilitates various optimizations. TAC requires that statements have at most three operands.

Compilers often take multiple optimization passes over each IR, and each pass may affect subsequent passes. The following are hardware-independent and hardware-dependent optimization passes common in the compilation of DL models (*italicized* passes are the most critical for performance in DL):

- **Hardware-independent optimizations**: *operator fusion*, loop permutations, arithmetic simplification, constant folding and propagation, dead code elimination, common subexpression elimination, inlining, loop-invariant code motion, and memory to register promotion.

- **Hardware-dependent optimizations**: *loop tiling, polyhedral transformations, data layout manipulations, operator folding, micro-kernel and intrinsic matching, memory allocation, memory fetch sharing, device placement, operator scheduling,* loop splitting, and loop fission.

Operator fusion and loop tiling are the most important optimizations for DL models, followed by the other italicized optimizations. Some operator fusions may be hardware-dependent; those that are ISA-dependent are encompassed under operator folding. All these optimization passes are discussed in Sections 8.4 and 8.5.

In the remainder of this chapter, we review programming language types. We explain the compilation process from high-level language to machine code and, as an example, explain how this process works with the popular LLVM compiler. Moreover, we describe standard compiler optimization passes to accelerate the execution of DL models. Specific DL compilers are discussed in Chapter 9.

8.1   LANGUAGE TYPES

Languages can be characterized as *statically-typed languages* or *dynamically-typed languages*. In a statically-typed language, the *variables* are associated with a data type that does not change. Statically-typed languages are generally compiled languages; the type checking happens at compile time before the program runs. Statically-typed languages include C/C++, CUDA C/C++, Java, Scala, Fortran, and Pascal.

In a dynamically-typed language, the *values* are associated with a type, and the variables can change type. That is, the variables are dynamic and can be thought of as generic pointers to typed values. Dynamically-typed languages are generally interpreted languages; the type checking happens at runtime. Dynamically-typed languages include Python, JavaScript, and PHP.

Languages are strongly-typed or weakly-typed. While there is no universally accepted distinction between them, in general, a strongly-typed language requires that every value has a type, and a variable must be explicitly cast before it is assigned to another variable of a different type.

8.2   FRONT-END, MIDDLE-END, AND BACK-END COMPILATION PHASES

Compilers, such as GCC, LLVM, ICC, MSVC, and some of the DL compilers discussed in the next chapter, lower code to a target ISA. The compilation process from a high-level language to machine code typically consists of three overarching phases, illustrated in Figure 8.1:

1. front-end compiler: *parser* (language dependent);

2. middle-end compiler: *optimizer* (language and hardware independent); and

Figure 8.1: The compilation process consists of a front-end, middle-end, and back-end phase.

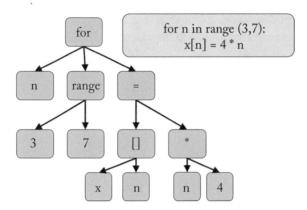

Figure 8.2: (green) The programmer's source code. (orange) The abstract syntax tree (AST) representation. The parser constructs an AST that captures the lexical structure of the source code.

3. back-end compiler: *machine code generator* (hardware dependent).

Each phase has one or more IRs depending on the optimization passes. One or multiple compilation infrastructures may be used for these phases.

Front-end The front-end compiler parses the code, converts it into tokens, checks for errors (syntactic and semantic analysis), and generates a domain-specific IR. Two common types of IR used by front-end compilers are the abstract syntax tree (AST) and the control-flow graph (CFG) data structures. The AST is language-dependent. It captures the lexical structure (layout) of the source code, using the internal nodes for the statements and operators, and the leaf nodes for the operands representing values or variables. The parser returns an error message if a rule in the language specification is violated. Front-end compiler algorithms are fairly matured. Figure 8.2 illustrates an AST generated from a `for` loop.

A CFG is language-independent and expresses the control-flow and data paths through a program. A control-flow statement, such as `for`, `while`, and `if`, determines which of two or more paths to take. The nodes are basic blocks, and the edges represent possible execution paths between basic blocks. Basic blocks are a set of sequential operations with no branch statements until the end of the block. Figure 8.3 illustrates a CFG used to compute the factorial of N. The top block is for the code that runs before the `while` loop. The next block is the comparison to

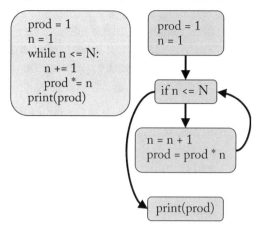

Figure 8.3: (green) The programmer's source code. (orange) The control-flow graph (CFG) representation. The CFG expresses the possible decisions at each graph node.

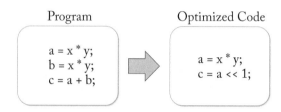

Figure 8.4: The optimizer reduces the number of operators that need to be executed: (left) the unoptimized code and (right) the equivalent optimized code assuming a is an unsigned integer.

decide which branch to take. The next block is the body and returns to the comparison. The last block is the code that runs after the while loop. A CFG is typically compiled from an AST IR.

Middle-end The middle-end compiler has two main tasks: (1) canonicalize the various ways of representing the code into predictable patterns removing unnecessary variations and (2) improve the performance via a series of optimizations. Some middle-end optimizations are completely hardware-agnostic, and others need information about the back-end hardware, such as multi-threaded parallelization and SIMD vectorization. Figure 8.4 illustrates an example optimizing the equation $c = a + b$ as $c = a << 1$, where the operator $<<$ left-shifts a by 1 bit, which is equivalent to multiplication by 2.

The optimizer typically performs a series of distinct optimization passes on the IR. LLVM does around 150 passes. GCC and LLVM use different algorithms to traverse the IR iteratively. While the order of optimizations affects the end result, strict rules to determine the optimal order do not exist.

In general, there are three common compiler optimization parts: legality analysis, profitability analysis, and transformation. Legality analysis makes sure the transformation does not break the program. Profitability analysis uses a cost model to determine if the optimization is beneficial and searches for parameters to perform the optimization. Finally, the transformation performs the actual modification of the code.

Back-end   The back-end compiler lowers the IR onto the target ISA and performs hardware-dependent optimizations. These include instruction selection, instruction scheduling, and memory and register allocation.

The output from the back-end compiler is machine code in an assembly file or object file. The linker takes the object file(s) and dependent libraries to generate an executable file.

Intrinsic functions   There are some constructs, such as vectorization with SIMD instructions, that a high-level language may not address. In these cases, intrinsic functions provide a way for the programmer to use such constructs. An intrinsic function is a function used in a given language. The implementation is handled especially by the compiler, which maps and optimizes the intrinsic function for a back-end target. Typically, the compiler substitutes a sequence of instructions for the intrinsic function call. Some intrinsic functions are portable, and others are target specific.

An intrinsic function provides a compromise between transparent integration inside a C/C++ function and writing full inline assembly (where most instructions map directly to an ISA instruction and the compiler takes care of register allocation). GCC, for instance, implements intrinsics for C/C++ that map directly to the x86 SIMD instructions.

8.3   LLVM

LLVM originally stood for *low-level virtual machine* (albeit with no relationship to what most current developers today think of as *virtual machines*) since the low-level LLVM IR code targets a universal theoretical machine (hence the original term *virtual*) and compiles for a variety of architectures [LA04]. While the concept is still accurate, LLVM is now the full name and no longer an acronym. LLVM is a brand for an umbrella project applied to the following:

- LLVM IR

- LLVM Core

- LLVM debugger

- LLVM implementation of the C++ standard library

- LLVM foundation

In this section, LLVM refers to the LLVM Core, a middle-end and back-end compiler program written in C++.

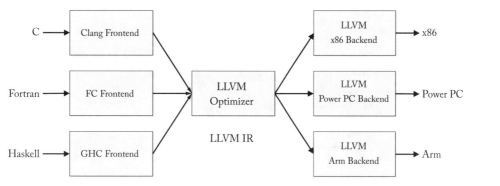

Figure 8.5: LLVM is designed as a set of modular compiler components supporting various front-end languages and back-end hardware targets.

LLVM is designed as a set of reusable libraries with well-defined interfaces. A front-end compiler can be written for any language that can compile to the LLVM IR, and a back-end compiler can be written for any new target that can compile from the LLVM IR, as illustrated in Figure 8.5.

8.3.1   FRONT-END COMPILERS

Clang is an "LLVM native" front-end compiler for the C/C++, Objective-C/C++, and CUDA C/C++ languages. Clang and LLVM are used in production to compile all the apps in Apple's iPhone and iOS, Google's production server applications, Nintendo GameCube games, and Sony's Playstation 4 games.

In addition to Clang, other front-end compilers to support multiple languages, such as Python, TensorFlow, Halide, Julia, Swift, and Fortran, can compile to LLVM. Thus, all those languages can exploit the same LLVM compilation infrastructure for compiler optimization passes and back-end machine code generation. The LLVM back-end compiler supports the x86, x86-64, PowerPC (PPC), Nvidia Parallel Thread Execution (PTX), AMD Graphics Core Next (GCN), Arm, and other architectures.

LLVM does not support domain-specific expressions and types. Therefore, domain-specific SSA-based IRs, such as those shown in Figure 8.6, should be used for optimizations that are too high-level for LLVM. A higher-level IR provides better location tracking to improve debuggability, traceability, and user experience. However, this comes at the expense of heavier infrastructure and some duplication in the domain-specific IR and LLVM IR optimizations.

Swift, Rust, Julia, and the TensorFlow library all use a higher-level IR. Swift uses the Swift High-Level IR (SIL), and TensorFlow uses XLA HLO before lowering to the LLVM IR. These higher-level IRs are similar to the LLVM IR but have domain-specific expressions and types; in particular, TensorFlow supports DL operators on tensors.

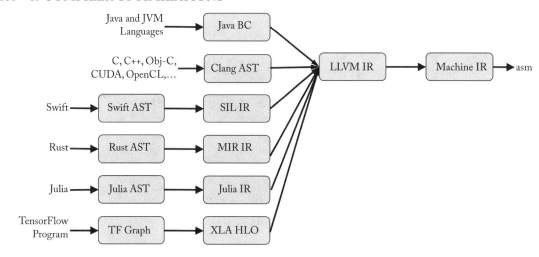

Figure 8.6: Many languages use a higher-level domain-specific IR for domain-specific optimizations before lowering to the LLVM IR. Based on [LP19].

8.3.2 INTERMEDIATE REPRESENTATION

The LLVM IR code is a self-contained (complete code representation), strictly SSA, strongly-typed, and mostly TAC language with well-defined semantics. It has three isomorphic forms: (1) an on-disk binary "bitcode" format (`*.bc`); (2) an assembly-like human readable/writable textual format (`*.ll`); and (3) an in-memory CFG data structure to which other front-ends can lower. LLVM provides tools to convert between these forms.

The LLVM IR has a simple architecture-independent instruction set made up of the following components: (1) operator instructions, (2) operands, (3) control-flow, and (4) `phi` nodes to manage the data flow. Additional concepts not discussed are intrinsics, metadata, and attributes. The following is a simple LLVM IR code sample that uses the first three of the four components mentioned above.

```
declare i32 @f(i32 %z)

define i32 @p(i32 %a, i32 %b) {
entry:
    %0 = mul i32 %a,%b
    %1 = call i32 @f(i32 %0)
    %2 = mul i32 %0, %1
    ret i32 %2
}
```

In line 1, the function `@f` with value `%z` is declared. In line 3, the function `@p` with integer arguments `%a` and `%b` is defined. `%0` equals the product of `%a` and `%b`; `%1` equals the returned value of function `@f` with argument `%0`; `%2` equals the product of `%0` and `%1`; and `%2` is returned value.

Figure 8.7: GCC can be used for the front-end, middle-end, and back-end compilation.

A `phi` node is an instruction used to merge multiple control-flow paths and multiple definitions of a variable selecting which definition to use. In the CFG, the `phi` instruction, when used, is always at the start of a basic block. The `phi` node has multiple pairs of operands; each pair consists of a value and a reference to a basic block. The basic blocks are the immediate predecessors to the basic block in which the `phi` instruction is located.

8.3.3   BACK-END COMPILERS

The optimized LLVM IR is passed to the LLVM back-end compiler for ISA code generation. The LLVM code generator provides many optimization passes by default. A programmer writing a new back-end compiler from the LLVM IR to a target architecture can use, overwrite, or add to the default passes. This flexibility allows the back-end compiler programmer to choose what makes sense for an architecture and reuse existing capabilities.

8.3.4   GCC COMPARISON

The most popular compilers are LLVM and the GNU compiler collection (GCC). GCC is a compiler program primarily written in C to compile various high-level languages to machine code. The GCC acronym has two different meanings: the *GNU C Compiler*, a C compiler, and the *GNU Compiler Collection*, a collection of compilers for the C/C++, Objective-C, Java, and Fortran programming languages. GCC began as a C compiler and evolved to support other languages. Today, GCC is the default compiler in Linux, and Clang/LLVM is the default compiler in macOS. A high-level comparison between GCC and LLVM follows:

- Performance: Relatively similar today (historically, GCC was faster).

- Modularity: LLVM is more modular.

- IR: LLVM IR is a complete code representation (although it is not intended to be used to write programs). GCC's IR (called GIMPLE) is not.

- Adoption: GCC has larger adoption; both have a large community of developers.

- License: GCC's GPL license requires developers who distribute extensions or modified versions of GCC to make their source code available unlike LLVM's Apache 2.0 license.

8.4   HARDWARE-INDEPENDENT OPTIMIZATIONS

The overarching goal of hardware-independent optimizations is to reduce memory accesses and reduce the number of operations. To that end, the following set of optimization passes are common. In DL, some of these optimizations are referred to as graph compilations, and the most important is operator fusion.

Operator fusion merges operators (also known as graph nodes) to reduce memory accesses by not having to save the intermediate results in memory. It is applicable when the operators have compatible loop patterns with continuous (called *coalesced* in GPU parlance) memory access. To illustrate, a fused sigmoid operator (see Figure 2.1) computes the exponentiation, addition, and division components keeping the intermediate results in local caches or registers and only saving the final result to memory.

Fused operators require that either the primitive libraries, such as oneDNN, MIOpen, and cuDNN, or that a back-end compiler provides or generates an optimized fused primitive to get the performance benefit. Thus, it is not entirely device-independent. Note that operator folding is a hardware-dependent operator fusion pass discussed in Section 8.5.

The types of operator fusions are:

- element-wise operator with another element-wise operator, for instance, the multiple element-wise operators in a sigmoid function;

- element-wise operator with a reduction operator, for instance, in the softmax function; and

- matrix-wise operator with an element-wise operator.

An example of the last bullet is a convolution or a GEMM operator fused with an activation function that operates on each element of the tensor, such as convolution followed by ReLU. The activation function is applied immediately after the output tensor value from the convolution is computed, and while this value is still in a register or scratchpad. Some of the fusion operators supported by TensorFlow's built-in compiler, Grappler (introduced in Section 9.2.7), are:

- Conv2D + BiasAdd + <Activation function>

- Conv2D + FusedBatchNorm + <Activation function>

- MatMul + BiasAdd + <Activation function>

- FusedBatchNorm + <Activation function>

As an example of the fusion benefits, Intel reported around 80× performance gain for batch size 1 fusing group convolutions in the MobileNet v1 model [SPE19]. In group convolution (introduced in Section 3.2.1), the different feature channels across a data batch are divided up

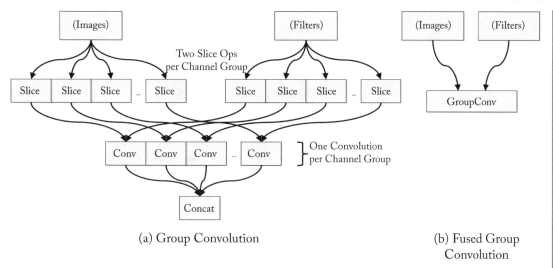

(a) Group Convolution

(b) Fused Group Convolution

Figure 8.8: (a) A group of convolutions used in MobileNet v1. (b) A fused operator can be jointly optimized for the entire group. Based on [SPE19].

into groups processed independently. The fused group convolution is jointly processed as a single DL operator, as shown in Figure 8.8.

Loop permutations modify loop indices to improve memory access. Some permutations, such as loop tiling, are target-dependent and are discussed in the next section. An example of permutation is interchanging `for` loops, as shown in the following code. The indices are interchanged to have coalesced memory accesses, which are faster than strided memory access.

```
// before loop permutations
for (i=0; i<N; i++)
  for (j=0; j<M; j++)
    x[j][i] = y[j][i]; // strided memory access
```

```
// after loop permutations
for (j=0; j<M; j++)
  for (i=0; i<N; i++)
    x[j][i] = y[j][i]; // coalesced memory access
```

Arithmetic simplifications reduces the number of expressions and simplifies the code. Examples include these replacements:

- $a \times x + b \times x + c \times x \Rightarrow (a + b + c) \times x$

- $!(x < y) \Rightarrow x \geq y$

- $2 \times x \Rightarrow x << 1$ (for unsigned integers)

- $x - x \Rightarrow 0$

- $x - 0 \Rightarrow x$

- $(x \times 2) - x \Rightarrow x.$

- $A^T B^T \Rightarrow (BA)^T$

- $(A^T B^T)^T \Rightarrow BA$

The last two items are known as **transpose eliminations**, which are a subset of arithmetic simplifications. Some of the simplifications can lead to numeric differences compared to the original expression. Still, these differences are generally small and can be safely ignored in DL.

During inference, the batch normalization expression can be incorporated into the convolution expression by scaling the weight values, as detailed in Section 2.6. While this is sometimes referred to as a fused operator, this optimization is an arithmetic simplification.

Constant propagation and **constant folding** substitute (propagate) known constants values in the expressions, and precompute (fold) constant expressions. Examples include these replacements:

- $3 \times 4 \Rightarrow 12$

- $x = 2;\ y = 3 \times x \Rightarrow y = 6.$

Dead code elimination (DCE) eliminates unused code. In the following code samples, the `if` expression is eliminated. Note that a has to be an integer (not a float).

```
// before constant propagation and DCE
int a=0;
if (a)
  mycode();
```

```
// after constant propagation
int a=0;
if (0)
  mycode();
```

```
// after DCE
int a=0;
```

Common subexpression elimination (CSE) eliminates repeated common subexpressions computing them only once. In the following example, the expression $a + b$ is only computed once, after the CSE pass.

```
// before CSE
c = a + b
d = a + b
e = c + d
```

```
// after CSE
c = a + b
d = c
e = c + d
```

Inlining, also known as inlining expansion, (not to be confused with the unrelated C++ *inline* keyword) moves the code of the called function into the calling function. It saves the overhead of procedure calls and allows further optimizations at the calling function at the expense of a larger executable file and, therefore, longer load time and increased pressure on the instruction cache. A toy example follows:

```
// before inlining
myFunction(int x){
  printf("%d\n", x);
  printf("%d\n", x*x);
}
myFunction(a);
myFunction(b);
```

```
// after inlining
printf("%d\n", a);
printf("%d\n", a*a);
printf("%d\n", b);
printf("%d\n", b*b);
```

Note that inlining wrapper functions do not affect the size of the executable.

Loop-invariant code motion (LICM), also called hoisting or scalar promotion, moves out expressions that are not required to be in the loop.

Memory to register promotion tries to promote memory references to be register references in order to reduce the number of memory loads and stores. The front-end and middle-end compilers assume an unlimited number of registers. Register assignment happens in the back-end compiler and is hardware-dependent.

8.5   HARDWARE-DEPENDENT OPTIMIZATIONS

The overarching goal of hardware-dependent optimizations is to have coalesced memory access and maximize data reuse (or equivalently, maximize arithmetic intensity). To that end, the following set of optimization passes are common. In DL, some of these optimizations are referred to as tensor compilations.

Loop tiling modifies the loop to improve the locality of memory access in space and time. It is one of the most impactful optimizations and one of the most challenging given the number of tile sizes (also known as stencils) and blocking strategies across the memory hierarchy. Cache blocking and register blocking use loop tiling and data layout optimization passes to maximize data reuse and minimize memory and register conflicts. These conflicts are known as bank conflicts and result when loading and storing data to and from the same location.

The goal of loop tiling optimization is to reuse data in local memory and reduce main memory accesses. This optimization tiles the loop so that the data in the inner loops fit into the local memory to increase reuse before it gets evicted. Loop strip-mining, also known as sectioning, is 1-dimensional tiling used for vectorization.

The code below (based on [VAK19]) demonstrates loop-tiling in two steps. In Step 1, the inner loop is split into two loops: an outer loop, known as the tiled (or blocked) loop, and an inner loop known as the element loop that fits into local memory. In Step 2, the outer two loops are permuted (swapped). This more efficient implementation reuses elements in the inner loop N times before they get evicted from the local memory. In the code, M is assumed to be a multiple of TILE. Note that the loop indexed by i could also be tiled (not shown in the code).

```
// Original code
for (i = 0; i < N; i++)
  for (j = 0; j < M; j++)
    operation(x[i], y[j]);
```

```
// Step 1: Tiling
for (i = 0; i < N; i++)
  for (jj = 0; jj < M; jj += TILE)
    for (j = jj; j < jj + TILE; j++)
      operation(x[i], y[j]);
```

```
// Step 2: Permuting
for (jj = 0; jj < M; jj += TILE)
  for (i = 0; i < N; i++)
    for (j = jj; j < jj + TILE; j++)
      operation(x[i], y[j]);
```

The optimal stencil (tile size) is unique to each microarchitecture and is a parameter the compiler has to select, adding complexity to the solution space. One algorithm to facilitate the selection is the Cache-Oblivious Recursion algorithm [FLP+99].

Polyhedral is a compiler technique that results in a set of loop transformations used for efficient code generation. Note that some of the polyhedral transformations are hardware-independent. A polyhedral representation specifies the boundary of a polyhedron (the index space of a tensor expression). The polyhedral-based compilations provide a set of (usually affine) loop transformations, such as loop tiling, to facilitate efficient code generation on a hardware target.

Polyhedral compilation techniques are conventional in HPC and image processing. The challenge is the NP-complete algorithms, such as integer linear programming (ILP) solvers or other exponential algorithms required, which limit scalability.

An affine representation is a simplified polyhedral representation with for loops and if control structure ops. An affine transformation applies a unique affine function to each element of a tensor and preserves the dimensions of the tensor. An affine compilation does not require the use of ILP or any other NP-complete algorithms. The DL compilers PlaidML, TVM, and MLIR dialects, such as LinAlg and Affine, use polyhedral-based (typically, affine-based) loop transformations. Chapter 9 covers these compilers.

Data layout, also known as memory format, memory layout, or tensor layout transformations, modifies the data layout so it is efficiently accessed. As reviewed in Section 2.3, standard data layouts used by the main frameworks are *NCHW* or *NHWC*, and *RSCK* or *KCRS* for the weight tensors. These data layouts are referred to as plain formats or native formats (native or default to the DL framework).

Data in memory is arranged as a 1D vector. The *NCHW* format means the width values are the innermost dimension and are adjacent in memory. The memory index offset for a given index $n, c, h, w \in N, C, H, W$ is

$$\texttt{offset}(n, c, h, w) = n \times CHW + c \times HW + h \times W + w.$$

TensorFlow and PyTorch natively support both *NCHW* and *NHWC* with *NCHW* as the default layout. ONNX only supports *NCHW*. FBGEMM and the Quantized Neural Networks PACKage (QNNPACK) support *NHWC* but not *NCHW*. LIBrary for eXtra Small Matrix Multiplies (LIBXSMM) supports both but is optimized for *NHWC*.

The data layout can be modified to achieve better reuse from cache (also known as local memory in AMD GPUs or shared memory in Nvidia GPUs), scratchpad, and registers to use SIMD, SIMT, or dataflow instructions more effectively. To illustrate, one of the layouts used by oneDNN for CPUs for architectures with 512-bit registers and $fp32$ values is the 5D tensor

$$N\hat{C}HW16\hat{c}.$$

This format blocks (tiles) the channel dimension in blocks of 16 to fit into a 512-bit (16 $fp32$ values) register. The memory index offset, using the $N\hat{C}HW16\hat{c}$ layout, is:

$$\texttt{offset}(n, c, h, w) = n \times CHW + \left\lfloor \frac{c}{16} \right\rfloor \times 16HW + h \times 16W + w \times 16 + (c \bmod 16),$$

where $\lfloor \cdot \rfloor$ is the floor operator. Using this layout format, the data is fed as 16 consecutive $fp32$ values into a register from the same n, h, w indices but different channels and processed in parallel using SIMD instructions. A channel size multiple of 16 is beneficial for this blocked format.

The cuDNN primitive library typically uses the *NCHW* layout. However, newer GPUs, such as the V100, prefer the *NHWC* layout for $fp16$ computations with C being a multiple of 8 to use the available tensor cores efficiently. Padding the channels with zeros to the desired size can improve the computational efficiency. Note that TensorRT supports blocked formats to achieve the highest performance on some workloads.

Depending on the operands, different layout strategies result in better performance. For instance, the convolution function potentially uses three different tensor layout strategies depending on the operand sizes:

1. one layout for operands with a small number of channels;

2. one layout for operands with a large number of input activations; and

3. one layout for operands with a large number of weights.

The *im2col* GEMM-based convolution, discussed in Section 2.3, can be used when none of the layout strategies are well optimized for the particular operand size. Similarly, the operands in a GEMM expression also use different layouts.

Layout optimizations involve inserting layout conversion nodes in the computation graph. The overall performance gains from the more efficient data layout should consider the data layout conversion time to determine if it has a net performance gain. A related graph-level optimization pass involves reducing the number of layout conversions by keeping the tensors in a blocked layout for as many nodes as possible before converting back to the default graph data layout.

Operator folding (ISA matching) combines two operators into one supported by a hardware instruction. For instance, the hardware may support a fused multiply-and-add (FMA) operator, in which case a fused operator, such as $a * b + c$, can speed up the program execution.

Micro-kernel and **intrinsic matching** matches and replaces a block of computations with the corresponding micro-kernel or corresponding hardware intrinsic.

Memory allocation, also known as static memory planning, preallocates runtime memory to hold the tensors and determines memory reuse and in-place computations.

Memory transfers, also known as explicit memory latency handling, is used in conjunction with the memory allocation pass. It adds memory access instructions to transfer data to and from memory banks to overlap memory transfers with computations. This optimization is even more critical for accelerators with limited control logic, as they have a limited ability to reduce memory access latencies or hide them within the computation. CPUs and GPUs have built-in mechanisms that hide the memory transfer latency, such as speculative prefetching. Some CPUs also have simultaneous multithreading (SMT), which keeps several threads persistent on a unit and interleave their executions to cover latencies.

Accelerators usually do not have caches but rather scratchpads, which need to be managed explicitly by software. Outsourcing memory control from the hardware to the software is a common ASIC design choice to reduce the size of the die and save power. For instance, the control logic of the TPU is small and only takes 2% of the die (the control logic in a GPU or a CPU is significantly larger) [SYP17].

Memory fetch sharing, also known as nested parallelism with cooperation, improves fetching data, from memory. Threads with local shared memory space cooperatively fetch data from higher levels in the memory hierarchy. GPU and accelerators with local shared memory benefit from this optimization.

Device placement evaluates the various back-end hardware targets and determines a strategy to allocate the execution of subgraphs to each available device.

Operator scheduling specifies which operators to schedule and their execution order, to reduce runtime memory usage and output latency.

Loop unrolling replaces the statements in the loop body with multiple statements to (1) reduce the number of loop control instructions, (2) improve parallelism if the statements are independent, and (3) reduce the branch penalty. Each iteration of the resulting loop executes multiple iterations of the original loop. This optimization typically increases the size of the binary. A toy example follows:

```
// Before loop unrolling
for (i = 0; i<100; i++)
  x[i] = i;
```

```
// After loop unrolling
for (i=0; i<100; i+=4){
  x[i] = i;
  x[i+1] = i + 1;
  x[i+2] = i + 2;
  x[i+3] = i + 3; }
```

When the number of loop iterations is not known until runtime, an AOT compiler can generate several versions of the loop with different unrolling factors, or alternatively, a JIT compiler can be used.

Loop splitting splits the loop iterations into multiple loops if the iterations are not dependent on each other and can execute in parallel. A toy example follows:

```
// Before loop splitting
for (i = 0; i<100; i++)
  printf( "Iteration %d\n" , i);
```

```
// After loop splitting
for (i = 0; i<25; i++)
  printf("Iteration %d\n", i);
for (i = 25; i<50; i++)
  printf("Iteration %d\n", i);
for (i = 50; i<75; i++)
  printf("Iteration %d\n", i);
for (i = 75; i<100; i++)
  printf("Iteration %d\n", i);
```

Loop fission, also known as loop distribution, splits the body of a loop if the components are not dependent on each other and can execute in parallel. Note that the reverse is called loop fusion which unites multiple loops into a single loop. To illustrate:

```
// Before loop fission
for (i = 0; i<100; i++){
  a[i] = 3 * i;
  b[i] = 4 * i; }
```

```
// After loop fission
for (i = 0; i<100; i++)
  a[i] = 3 * i;
for (i = 0; i<100; i++)
  b[i] = 4 * i;
```

In this chapter, we review the basics of programming languages and compilers that map high-level languages to machine code. We highlighted standard compiler optimization passes to accelerate the execution of DL models, particularly fusing element-wise operations into dense linear operations. Compilers are imperative for the success of dedicated DL processors; manually optimizing a model to perform well on a back-end target is extremely costly and not scalable across several targets. In the next chapter, we discussed prominent DL compilers used by hardware vendors and hyperscalers.

CHAPTER 9

Frameworks and Compilers

A framework has multiple types of compilers: the computation graph optimizer, the primitive libraries JIT to select the best schedule, the code generation path for operations not supported by the primitive libraries, and the other DL compilers covered in Sections 9.4–9.9. A computation graph is a high-level graph that represents the computations, data flow, and control-flow of a DL program (a model). Each node typically corresponds to a tensor operator (op), such as conv2d, GEMM, or sigmoid. Some nodes represent control-flow operators to enable conditional branches and loops on different parts of the computation graph. The edges represent the data flow and tensor dependencies between operators, as illustrated in Figure 1.5. A tensor is a multidimensional array; a scalar, vector, and matrix are 0D, 1D, and 2D tensors, respectively. 3D, 4D, and 5D tensors are also common.

In the DL compiler literature, the terms *expression*, *primitive function*, and *compute kernel* are often used interchangeably, with kernel primarily used as a synonym for convolutional filter. The order of transformations and other operations to execute a primitive is called a *schedule*. An expression consists of operands and operators. The operands are the tensor inputs and outputs of the primitive function. An example is the sigmoid expression composed of a division, an addition, and an exponentiation operator: $\sigma(x) = \frac{1}{1+e^{-x}}$. Given the frequent use of this particular expression or pattern of operators, it is beneficial to fuse them into a single sigmoid operator to reduce memory accesses, as explained in Section 8.4.

Frameworks, such as TensorFlow and PyTorch, implement over a thousand of operators for x86 CPUs and Nvidia GPUs. TensorFlow and PyTorch have built-in graph optimizers (Grappler in TensorFlow; JIT in PyTorch (no fancy name)) and a scheduler to execute the computation graphs. The scheduler (also known as execution runtime) dispatches each tensor operation to a precompiled target-specific implementation available in a primitive library if the library supports the operator. Frameworks also have a code generation path to supplement these libraries, such as through LLVM.

Low-level libraries, such as oneDNN, cuDNN, ROCm MIOpen, Eigen, OpenBLAS, GotoBLAS, and BLIS, provide optimizations to standard DL or basic Math functions (such as those introduced in Chapter 2) for a particular ISA. Intel and Nvidia work with the framework maintainers to co-optimize the framework's built-in graph optimizer and integrate the APIs of their respective primitive libraries.

Inference engines (IE) are inference-only frameworks. They are used to optimize and deploy already-trained models. IEs are leaner than the main frameworks focusing on inference-only graph-optimizations. While the frameworks have mostly consolidated to TensorFlow (TF) and PyTorch, the number of inference engines is slowly growing. Some of the prominent IEs are as follows:

- TensorFlow Lite for Android, iOS, and Raspberry Pi devices.

- TensorFlow.js for JavaScript environments, such as a web browser and Node.js.

- OpenVINO for x86 CPUs, and Intel's GPUs, FPGAs, and Movidius VPUs (used for edge devices).

- TensorRT for Nvidia's GPUs.

- ONNX Runtime (ONNX RT) for ONNX models, used by Microsoft and gaining adoption elsewhere. It uses OpenVINO, TensorRT, MLAS, and other back-end libraries. Microsoft is expanding ONNX RT to support training as a main framework.

- AWS Neuron for TensorFlow, PyTorch, and MXNet models running on AWS Inferentia-based instances.

- Neural Magic for x86 CPUs.

- TensorFlow-Serving for large-scale data center deployment of TF models. TF-Serving provides a layer over TF to handle inference requests over networks, and uses TF for the inference, so in this respect is not faster or leaner than TF.

- TorchServe for efficient end-to-end inference of PyTorch models. TorchServe provides a layer over PyTorch, similar to TF-Serving with TF.

A limitation of most IEs and ONNX models is that they do not provide all the inference functionalities and operators that the frameworks have. To mitigate this, most IEs have mechanisms to extend their functionality. Some IEs are integrated with a framework so that operators not supported by the IE fall back to the framework. For instance, TensorRT is integrated as a back-end runtime executor in TensorFlow. However, this increases the size of the inference software package and may not be practical in edge devices.

The current approach to train and deploy models is to use frameworks or inference engines with built-in graph optimizers and rely on libraries for target-dependent optimizations to standard primitives. The combination of F frameworks, M microarchitectures (hardware targets), P primitives, S schedules per primitive (a primitive function may have different schedules depending on the size of the operands), and D different numerical formats has an implementation cost in the order of $O(FMPSD)$.

This engineering approach is not scalable. It is a nontrivial engineering effort to optimize each combination. Also, the optimizations have to be upstreamed into each of the popular frameworks or the primitive libraries. Even standard primitives, such as convolution and LSTM, can have different optimization strategies. For instance, the optimal schedule for convolution depends on the batch size, the filter size, the number of channels, and the algorithmic implementation, such as direct, FFT, or Winograd (introduced in Section 2.3). Dispatching between these algorithms to pick the best performing one for a particular shape is a nontrivial task. Also, LSTM units can have different variants, each one requiring a unique optimization strategy.

An additional challenge with the current approach is that a graph optimization may suggest fusing two primitives to reduce memory access overhead. However, the fused primitive may not be optimized in the primitive library. The challenge to engineer optimized primitives is compounded by (1) the wave of DL hardware being developed, each with a unique microarchitecture, computational paradigm, and memory hierarchy and (2) the heterogeneous design of modern devices. A single device, such as a modern mobile phone, may have a CPU, GPU, DSP, ASIC (for matrix multiplications), and IPU (for image and video decoding).

DL system engineers at hyperscalers typically write code across all levels of the software stack. The cost of adopting a new hardware architecture may be prohibitive if it requires rewriting the code to run on the new hardware and, worse, if it requires learning a new programming language. There is a market to automate whole-program optimizations to reduce this cost, and startup companies, such as OctoML, Neural Magic, DeepCube, and OnSpecta, are offering optimizations as a service.

The overarching goal of graph compilers is to automatically generate efficient code for a diverse set of workloads and across different hardware targets that execute at similar or better efficiency to one manually optimized. The compiler lowers a model from the framework representation to a canonicalized high-level domain-specific IR. It then performs a series of target-independent optimizations passes, such as operator fusion, and eliminating unnecessary expressions (as discussed in Section 8.4). The optimized IR is functionally equivalent to the IR before the optimization passes. The compiler then lowers the optimized high-level IR to a low-level IR with limited to no control-flow and performs a series of target-dependent optimization passes, such as additional fusion passes, and data layout transformations. These passes aim to use the memory hierarchy and compute available efficiently. The compiler then either generates executable code for the back-end target or passes the optimized low-level IR to a target-specific compiler, such as an LLVM back-end compiler, to generate executable code.

The main challenge compiling expressions (primitives) is to constrain the space of functionally-equivalent schedules for each expression for a particular hardware target so that an efficient schedule for each expression is quickly selected. The space of schedules is enormous for each expression and hardware target pair. Finding the optimal schedule is an NP-complete problem with potentially billions of choices for a single expression. RL and ML techniques can improve DL compilers. One example is AutoTVM, discussed in Section 9.4. Note that passing contextual information from the high-level IR to a low-level IR can limit the search space. However, the low-level IR typically does not contain contextual information.

Another challenge is optimizing dynamic graphs, which are graphs with arbitrary allocations (variable input lengths), control-flows, iterations, and recursions [YAB+18]. Static computation graphs work well for directed acyclic graph (DAG) models with a fixed number of layers and dimensions per layer, such as ResNet-50. However, modern NLP, computer vision, and RL models, as well as recursive NNs and GNNs, have dynamic graphs with variable length inputs.

Most graph compilers have limited support for dynamic models. There are three primary methodologies to represent and use dynamic models with dynamic inputs and control-flow. First, compiling and caching multiple static graphs for different inputs, and selecting the best static graph at runtime. The programmer can provide lower and upper bounds on the input parameters to reduce the number of static graphs generated. Second, eliminating control-flow by unrolling loops or duplicating nodes at the expense of increasing the program's size and, hence, the memory requirements. Third, executing the control-flow in the (slower) Python interpreter, such as with PyTorch imperative mode (note that PyTorch JIT generates control-flow and data flow for declarative mode).

Frameworks and compilers should support these features:

- High- and low-level optimizations that are reusable across front-end frameworks and back-end targets.

- Strongly-typed tensors; that is, tensors with a known static shape and element type, such as *fp*32, *fp*16, *bf*16, *s*8, *u*8, or *bool*.

- Common tensor expressions, such as matrix multiplications, convolutions, element-wise activations, broadcast, reductions, and index transformations (such as reverse and concatenation).

- Fallback on the default implementation for subgraphs or expressions not supported by the compiler.

- Static graphs with placeholders for the tensor operands.

- Dynamic shapes.

- AOT and low-overhead JIT compilation (such as JIT without LLVM).

- Runtime executor for parallelism to concurrently execute subgraphs across threads or devices.

- Auto-differentiation and mathematical optimizers, such as SGD, for training.

- Collective communication primitives (AllReduce, AllGather, AllToAll, Broadcast) for distributed training.

Table 9.1: Features across various DL compilers

	TVM	GLOW	XLA	PlaidML
Auto-differentiation		✓		
Quantization		✓		
Dynamic Shapes				
Build-in control-flow	✓			
Target-independent HLO	✓	✓	✓	
Target-independent LLO	✓	✓	✓	✓
Target-dependent optimizations	✓			✓
AOT	✓	✓	✓	
JIT				✓
C or C++	✓	✓	✓	✓

The most prominent DL compilers (outside of the frameworks' built-in graph optimizers) are TVM, XLA, Glow, PlaidML, and various MLIR dialects (MLIR is a compiler infrastructure that supports various IRs or dialects and compiler passes). These compilers are written in C/C++ for speed and portability. While TVM is the most mature compiler today, all compilers are still in their infancy and have limited adoption in industry. This is likely to change in the next few years with the wave of DL hardware starting to hit the market, which increases the market demand for robust compilers. Table 9.1 provides a summary of key features from each of the main DL compilers outside the default framework compilers. Other less prevalent compilers are taco, Tensor Comprehension, DLVM, Weld, and Diesel. Sections 9.4–9.9 discusses these compilers and their adoption in industry.

While DL compilers aim to support multiple front-end frameworks, they are often developed by a team related to an existing framework that firstly focuses on that framework. In particular, XLA and MLIR dialects with TensorFlow, Glow with PyTorch, and TVM with MXNet. Nevertheless, compilers are expanding their front-end support.

Grappler (TensorFlow's built-in graph optimizer), PyTorch JIT, XLA HLO, and Glow compilers strive to optimize the inefficiency brought by the user program via target-independent optimizations. They rely on a primitive library (such as cuDNN, MIOpen, oneDNN, or Eigen) or another compiler for target-dependent optimizations. PlaidML, various MLIR dialects, and TVM support target-independent and dependent optimizations and back-end code-generation.

In this reminder of this chapter, we review the DL frameworks with a particular focus on TensorFlow and PyTorch, which have built-in graph optimizers and schedulers to execute the computation graphs. We also describe in more detail the prevalent DL compilers.

9.1   FRAMEWORKS

DL libraries or frameworks provide the programmer tools to define, train, and deploy models. Frameworks abstract many of the mathematical and implementation details. For instance, they contain functions or modules to differentiate a model with respect to a cost function (compute its gradients), so the programmer does not have to code the gradient computations. While the computational performance across the frameworks varies depending on the optimization techniques exploited, the statistical performance of the models trained across frameworks is essentially the same; they implement essentially the same mathematical algorithms.

Frameworks compile the program to a graph and optimize the graph. The nodes are implemented using C++, CUDA, or using a precompiled target-specific implementation available in a primitive library. Frameworks may also use a DL compiler to improve execution efficiency.

The most popular frameworks are TensorFlow developed by Google and PyTorch developed by Facebook, both written in C++ and have a Python wrapper. TensorFlow is the most popular framework in the industry and the second most popular in academia. PyTorch is the most popular framework in academia, the second most popular in the industry, and the fastest-growing framework [Lor19]. Other frameworks used in industry but (based on Google Trends) with limited adoption outside the companies that developed them are Apache MXNet, PaddlePaddle, and Flax/JAX. Amazon (in collaboration with the University of Washington, Carnegie Mellon University) developed MXNet, Baidu developed PaddlePaddle, and Google developed Flax/JAX (primarily for research). Flax provides high-level functions on top of JAX, a JIT compiler that uses Autograd and XLA for differentiation and executes NumPy code on CPUs, TPUs, and GPUs [Jax20]. NumPy is a library for Python for multidimensional tensor operations.

TensorFlow and PyTorch offer two programming paradigms: imperative programming and declarative (symbolic) programming. Imperative programming performs the computations as they run, and declarative programs separate the definition of the various expressions in the program from the execution. Gluon and the standard front-end MXNet, respectively, also adopt these paradigms.

In the remainder of this section, we provide a brief history and adoption of various frameworks. We discuss imperative and declarative programming styles and their tradeoffs as well as dynamic and static programming.

9.1.1   BRIEF HISTORY AND ADOPTION

A framework's popularity is one indicator of its usability and community support, including the number of tutorials, available GitHub repositories, and discussion forums. Over the past years, many frameworks have been developed (most are no longer actively maintained), such as (in alphabetical order) BigDL, Caffe, Caffe2, Chainer, CNTK, CoreML, CXXNET, Dark-Net, DL4J, DSSTNE, DyNet, fast.ai, IDLF, JAX/Flax, Kaldi, Keras, Lasagne, MXNet/Gluon, neon, ONNX RT, PaddlePaddle, PyTorch, Singa, TensorFlow, Theano, and Torch.

Theano was developed by the Montreal Institute for Learning Algorithms (MILA) and was the first widely adopted framework in academia. Caffe was later developed by UC Berkeley and gained rapid adoption in the market, primarily for computer vision tasks. For a couple of years, Caffe was the most popular framework in the industry. TensorFlow and Keras (a wrapper around TensorFlow and other now-deprecated frameworks) took over the top spot soon after TensorFlow was released. Theano and Caffe are no longer supported. Kaldi is a C++ toolkit for speech recognition research (supports DL and ML algorithms) and has enjoyed some success in that community for several years.

Other hyperscalers have attempted to replicate the success of TensorFlow: Microsoft with CNTK and later with ONNX RT, Amazon with DSSTNE and later with MXNet/Gluon, and Baidu with PaddlePaddle. The only other widespread success has been Facebook with PyTorch. Today, the market has mostly consolidated to TensorFlow and PyTorch, with MXNet, PaddlePaddle, and ONNX RT as a distant third, forth, and fifth, respectively.

9.1.2   PROGRAMMING STYLE: IMPERATIVE AND DECLARATIVE

ML libraries offer two programming styles: imperative where expressions execute immediately, and declarative where programmers first construct the dataflow graph, and then the entire graph is optimized and executed. TensorFlow v1 (there was no v0) was designed with a declarative programming style and PyTorch v0 with an imperative programming style. TensorFlow v2 and PyTorch v1 provide both styles to the programmer.

An imperative program performs each computation as the program executes. This is referred to as *define-by-run* or dynamic graph because the model or graph is defined dynamically as it runs, or as *eager execution* because the program is eager to compute and evaluate an expression as soon as the programmer requests to examine the output. For reference, most code written in Python is imperative.

Chainer adopted imperative programming in 2015. PyTorch v0, TensorFlow v2, and Gluon later followed Chainer's approach. The (imperative) code to implement a function across these frameworks has similar characteristics.

A declarative program first defines a function abstractly (symbolically) with no actual numerical computations. Then the program compiles and runs using actual tensor values. This is referred to as *graph programming* or *define-and-run* because the graph is first defined and then executed, or as *lazy execution* because the program waits until the graph is constructed and executes only the dependent expression, or as *symbolic programming* because the variables are symbols with no numerical values until it executes.

Declarative programming enables the AOT compiler to do analysis and optimizations across the entire set of expressions. The runtime exploits buffer reuse and inter-op parallelism. A single static computation graph often represents the entire topology.

A static graph is suitable for production environments to get maximum performance in terms of speed and memory. Having the entire computation graph available provides the compiler information on what values to compute (and ignores graph nodes that are not necessary for the desired output), how much memory to allocate, and what subgraphs to execute in parallel.

Declarative programs follow a dataflow programming model, that is, the program is represented as a computational (symbolic) directed graph with nodes representing computations and edges representing data. Special nodes in the graph represent the control-flow.

Imperative programs implement control-flow in the native language, which facilitates specifying arbitrary control-flow in the program. For instance, in PyTorch, the Python native control-flow statements if, for, and while are used. In addition to being more intuitive, this explicit control-flow facilitates using models with complex dynamic graphs (non-DAGs) often used in NLP and RL research.

9.1.3 AUTOMATIC DIFFERENTIATION

A step in training is computing the gradient of the cost with respect to each weight and activation via backpropagation. Once a model and a cost function are defined, the frameworks determine the computations necessary for the backpropagation. This feature is native in all popular frameworks and known as *automatic differentiation* (AD).

AD computes the gradient of one node with respect to the previous node and uses the chain rule to compute the gradient of the cost with respect to each weight and activation. In symbolic programming, this is done by adding gradient computation nodes. In imperative programming, this is done by recording or tracing the flow of values as they occur and generating a dynamic graph; the programmer marks the layers needed for AD.

9.2 TENSORFLOW

TensorFlow is an open-source library, written in C++, developed by Google with several contributors outside of Google. It was released in November 2015 and has become the most popular framework in the industry. It supports over a thousand different operators [SL19]. In addition to Python, TensorFlow supports other language APIs (some maintained by the broader community at various degrees of support), including Swift, Julia, C++, Scala, Java, JavaScript, Rust, and Go. Models trained by TensorFlow can deploy across various inference engines.

TensorFlow v1 is designed as a declarative programming style library [ABC+16]. Programmers construct an AST (the graph), usually in Python using a low-level API, and then compile and interact with the graph using a TensorFlow *session*. However, this low-level API has a steep learning curve and does not let the programmer use native Python control-flow or debuggers. TensorFlow v1 uses control-flow nodes, such as loop condition, switch, and merge nodes to represent data flow, which increases the complexity of pattern matching required for optimizations [YAB+18]. To facilitate v1 usage, higher-level libraries and APIs were developed, such as TFLearn, Slim, SKflow, and Keras. TensorFlow v1 is under maintenance mode, and all new work is going into TensorFlow v2.

The most notable changes from TensorFlow v1 to v2 are: (1) the Keras APIs are default, (2) eager execution is default, and (3) improved organization for APIs, functions, and namespaces. TensorFlow provides a conversion tool to port the code from v1 to v2. To help determine whether an online document or code sample refers to v1 or v2, note that v1 uses the following objects not present in v2: `tf.enable_eager_execution`, `session.run`, `tf.placeholder`, and `feed_dict`.

The remainder of this section is as follows: We introduce the Keras APIs infrastructure, the Estimator API, and the tools to convert a dynamic graph constructed in Eager-style code to a static graph using `@tf.function` and AutoGraph. We highlight the tools for distributed training, the TensorBoard visualization tool, the Profiler tool, and the compilation TensorFlow infrastructure. Other TensorFlow libraries and tools with some adoption in industry are TensorFlow Hub, TensorFlow Extended (TFX), TensorFlow Lite (TFLite), and TensorFlow Probability (TFP). TensorFlow Hub provides an extensive service of prebuilt models; end-users can fine-tune them or use them as preprocessing layers (such as some of the embeddings available). TFX is an end-to-end series of connected libraries use to deploy DL pipelines; specifically, TFX provides the critical parts of the DL pipeline except for the model building and training (which is core TensorFlow). TFLite is a lite framework for on-device inference. TFP is a library for probabilistic reasoning and statistical analysis.

9.2.1   KERAS APIS

In TensorFlow v2, Keras is part of TensorFlow (no need for a separate installation) and the default API. Keras is a widely adopted high-level API for defining and training models. Note that Keras has a reference implementation maintained as a separate project. TensorFlow includes a complete implementation of the Keras API (in the `tf.keras` module) with some enhancements.

A Keras model (and its superclass, `tf.Module`) is a way to store, access, and save variables. Keras is more opinionated than `tf.Module` about functionality; it provides abundant built-in support for ML/DL workflows. `tf.Module`, by contrast, is lightweight and unopinionated; it is used as the base class for Sonnet, DeepMind's high-level API built on top of TensorFlow.

TensorFlow v2 provides the Keras Sequential API and Keras Functional API for declarative programming, and the Keras Subclassing API for imperative programming. The styles are interoperable, allowing the programmer to mix and match. The Keras API abstracts many of the complexities of the low-level APIs, facilitating prototyping and debugging models. Note that the Sequential and Functional APIs may feel imperative, and developers may not realize they are using a symbolic programming style. Note that Keras models, saved with the SavedModel format, contain (among other things) a graphdef and weights, so they serve the same function as the saved graphs in TensorFlow v1 for most uses. Table 9.2 shows the pros and cons of these APIs.

The Keras Sequential API is used when the model does not have branches, and each layer has one tensor input and one tensor output. The Keras Functional API is used for directed acyclic

Table 9.2: Pros and cons of the programming styles offered in TensorFlow v2

API	PROs	CONs
Low-level	Maximum control	Difficult to debug Not pythonic Steep learning curve
Sequential	Simplest API	Sequential models only
Functional	Simple API Widely adopted	Static graphs only
Subclassing	Pythonic High flexibility Dynamic and static graphs `@tf.function` (Section 9.2.3) Similar to PyTorch API	Slightly more complex API

graphs (DAGs), where each layer may have multiple tensor inputs or outputs, shared layers, or nonsequential data flow, such as in residual connections. The Keras Subclassing API is used for imperative programming; the programmer defines a new class that inherits and extends the Keras Model class defined by the framework. This class imperatively defines a function with the model and a function with the forward pass (the backward pass is generated automatically). The low-level API from TensorFlow v1 is still available to use in TensorFlow v2.

We recommend using the Keras Subclassing API as it provides flexibility to develop and experiment with any type of model, including dynamic models. Also, it has a similar programming style to PyTorch, which can facilitate using both frameworks (it is not uncommon for different engineers in the same company to use one or the other).

9.2.2   EAGER EXECUTION

In TensorFlow v2, eager execution (rather than graph execution) is enabled by default [AMP+19]. Eager execution mode is primarily used to design and debug models, while static graph execution is used to deploy models in a production environment. The expressions are evaluated without initializing a session, unlike in TensorFlow v1 and `tf.Tensor` objects reference actual values rather than symbolic ones. This type of execution facilitates prototyping using Python control-flows (simplifying the design of dynamic models) and debugging with standard Python tools. In eager execution, `tf.GradientTape` records operations for automatic differentiation. Unless the programmer explicitly turns off `tf.GradientTape`, any op with variables gets automatically traced. The programmer only needs to watch static tensors that get passed in explicitly. In Keras, the programmer can mark layers as trainable (default) or not trainable.

9.2.3   `@TF.FUNCTION` AND AUTOGRAPH

A concern with eager execution is the lack of graph-level optimizations. To mitigate this, programmers can add the `@tf.function` decorator to their functions after successfully prototyping and before deploying it [AMP+19]. With `tf.function`, the backward pass is precomputed while tracing, so the gradient tape is not invoked. The `@tf.function` decorator translates Python programs into TensorFlow graphs for performance optimizations. A function called from an annotated function runs in graph mode and does not need to be explicitly decorated.

The `@tf.function` decorator is a JIT tracer. When the decorated function runs, it generates a graph function. If the tensor inputs change, a new trace of the Python function is triggered to generate a new graph function. These graph functions are polymorphic in their inputs; a single Python function can generate multiple graph functions.

This graph function represents Python control-flow internally as TensorFlow control-flow nodes using AutoGraph. AutoGraph is a feature of `@tf.function` that converts various Python constructs, such as `for`, `while`, and `if`, into TensorFlow graph equivalents, such as `tf.while_loop` and `tf.cond`. This conversion is required to port the graph to environments without a Python interpreter. AutoGraph supports arbitrary nestings of control-flow and can be used with complex models. Autograph also supports `assert -> tf.assert`, and `print -> tf.print`, gated on whether the arguments are tensors or Python objects.

9.2.4   ESTIMATOR

TensorFlow v2 keeps the Estimator API (including premade Estimators), another high-level TensorFlow API introduced in v1. Premade Estimators provide preimplemented, ready-to-use model functions for training and inference, such as Linear Classifier, DNN Classifier, Combined DNN Linear Classifier (Wide & Deep models), and Gradient Boosted Trees. Note, however, that using the Keras API is recommended over Estimators.

In `distribute.Strategy` in TensorFlow v2, the distribution toolkit was rewritten to build on the low-level parts of the library. Likewise, `tf.data`'s distributed-by-default approach in v2 makes a lot of the metaprograming in Estimators unnecessary.

9.2.5   TENSORBOARD

TensorBoard displays the graph, embeddings, and tensor distributions. It plots cost values during a run, which helps determine convergence and facilitates debugging. TensorBoard also compares various models and costs across training runs. In addition, TensorFlow enables the programmer to visualize the graph using `keras.utils.plot_model`, and `model.summary()` to get the description of the layers, weights, and shapes.

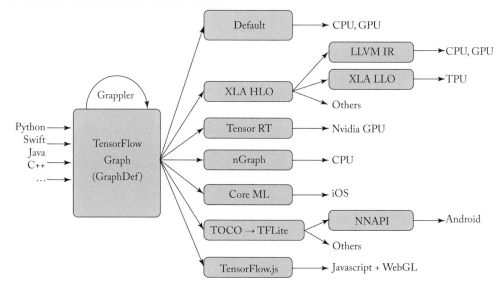

Figure 9.1: The TensorFlow IR GraphDef is optimized by Grappler and passed to other compilers for additional optimizations. Based on [Goo19].

9.2.6   PROFILER

Profiler tracks the performance of models and hardware consumption (time and memory) for the various operators. It can be used during training and inference to resolve performance bottlenecks and improve a model's performance on a CPU or GPU.

9.2.7   TENSORFLOW RUNTIME EXECUTION

The TensorFlow runtime is a cross-platform module that controls the execution of the graph. The TensorFlow code representing a computation graph is serialized to GraphDef format, the TensorFlow IR, using Google Protocol Buffers. During runtime, this serialized graph is optimized through the Grappler module, a device-independent graph optimizer to canonicalize and, depending on the back-end compiler, to optimize the graph. The TensorFlow runtime takes subgraphs and marks them for dispatch to its runtime or a different back-end runtime selected by the programmer, as shown in Figure 9.1. When an operator is not supported by the different runtime, it falls back to the default runtime.

During runtime, Grappler's MetaOptimizer provides high-level graph optimization. Some optimizations have a dependency on whether a node is supported by the primitive libraries. The runtime uses a LUT to find precompiled primitives from libraries, such as oneDNN and Eigen for CPUs, and cuDNN for Nvidia GPUs. The runtime also allocates memory and thread pools so parts of the graph can run in parallel.

Google is developing TFRT, a new TensorFlow Runtime, to replace the existing runtime and provide a unified, extensible infrastructure across various hardware targets. TFRT executes the optimized, target-specific kernels generated by MLIR, or directly uses (in eager execution) the TensorFlow API calls. It is unknown when Google will make TFRT the default runtime. We recommend hardware targeting market deployment in 2022 and beyond use TFRT.

9.3   PYTORCH

PyTorch is an open-source Python library for tensor computations similar to NumPy but with GPU support. It has built-in automatic differentiation and APIs for training and inference applications. PyTorch is maintained by Facebook with multiple contributors outside of Facebook. It was released in October 2016. It is the most popular framework in academia, the second most popular framework in the industry, and the fastest-growing framework [Lor19].

PyTorch v0 was designed as an imperative programming style library to facilitate research and development. For production-scale where performance is critical, Facebook developed the open-source Caffe2 graph-based execution library in April 2017. Facebook's servers and mobile app used Caffe2. To better interface between PyTorch v0, Caffe2, and other frameworks, Facebook partnered with Microsoft and later with other companies to develop the Open Neural Network Exchange (ONNX) format released in Sep. 2017. ONNX provides a standard format for various frameworks to exchange (export and import) extensible computation graph models for inference and, thus, streamline the path from research and development to production. A model would be developed and trained in PyTorch v0, exported to ONNX, and then imported into Caffe2 for production at scale.

PyTorch v1 (released in December 2018), hereafter referred to as just PyTorch, merges PyTorch v0 and Caffe2. PyTorch enables switching models from eager (imperative) mode to graph execution (declarative) mode, which further streamlines the path from research and development to production. Programmers develop, debug, and test their models in eager mode. They then migrate the models to graph mode for graph optimizations and may export a non-Python representation for scaled production in servers, mobile, or other platforms. Other key additions to PyTorch are a C++ API, JIT compilation, and a distributed library across Python and C++ environments.

PyTorch computation graphs are dynamic. PyTorch keeps track of the operators performed and builds a computation graph behind the scenes. Every time the programmer adds a layer, PyTorch rebuilds the computation graph. Automatic differentiation uses this computation graph.

PyTorch GPU expressions execute asynchronously, meaning the expressions can run in the GPU and synchronize with the CPU host when necessary, such as when copying data between host and device, or between devices. This synchronization is invisible to the programmer. For debugging, it may be useful to force synchronize-execution to trace an error.

PyTorch supports x86/64, Arm, and POWER CPUs and Nvidia GPU back-end targets. Support for other platforms is available via Glow. Google and Facebook added a PyTorch front-end to XLA to enable PyTorch programs to run on TPUs [She18].

9.3.1   PROMINENT PACKAGES

Some of the most prominent packages used in PyTorch are the following:

torch contains data structures and operators that operate on torch tensors. A torch tensor is a multidimensional array with all its elements being a single data type, such as $fp64$, $fp32$, $fp16$, $int64$, $int32$, $int16$, $int8$, or $boolean$.

torch.tensor is used to define and initialize torch tensors similar to NumPy `ndarray`. Note that there are various ways to convert from a torch tensor to a NumPy ndarray.

torch.nn contains the building blocks to models, similar to the Keras API. The `torch.nn.Module` is the base class for all NN modules. The program inherits this class to define a model. The `nn.Sequential` constructor adds the modules in the order they are passed, similar to the Sequential Keras API. The `nn.Module` is similar to the Subclassing Keras API.

torch.autograd is used for automatic differentiation. The programmer takes advantage of this package by marking which tensors should have their gradients computed. The gradients are computed (using the chain rule) when the `backward()` function is called on a variable. The inputs and labels have `requires_grad` set to `False` as those variables are not differentiable. Note that automatic differentiation is not part of the tensor library ATen (discussed in Section 9.3.2), but rather an augmentation on top of ATen.

torch.jit is the key differentiator from PyTorch v0. This package enables the seamless transition from eager mode to graph (also known as script) mode providing both flexibility and speed. It compiles the code (the annotated parts for compilation) to a statically typed graph representation and performs graph-level optimizations.

PyTorch provides two JIT modes, Tracing and TorchScript, shown in Figure 9.2. The `torch.jit.trace` mode works for models with no control-flow, such as the VGG model. The `torch.jit.script` mode is a statically-typed subset of Python that uses the TorchScript compiler. TorchScript translates the Python AST to a static graph. It is popular for models where control-flow is important, such as RNN models. A model may use both modes; in particular, when needing control-flow in a complex model, a scripted function can use a traced function on the portions of the model with no control-flow.

The `@torch.jit.script` decorator scrips a function or an `nn.Module`. The script mode can execute without Python. Using the C++ native module LibTorch, a Python-based model can be loaded and run in C++, which is useful for non-Pythonic environments, such as embedded systems.

torch.optim provides mathematical optimization algorithms to train a model, such as SGD.

torch.cuda provides CUDA tensors that utilize GPUs for computation.

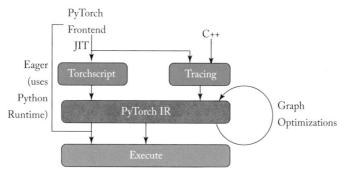

Figure 9.2: PyTorch can be executed in Eager mode via the Python runtime or in JIT mode via TorchScript, Tracing or both to generate a complete graph representation. This graph is optimized and then executed. Each expression is executed with the ATen library.

torch.distributed supports distributed training across multiple nodes using NCCL for GPUs and Gloo or MPI for CPUs.

torch.utils supports data loading and TensorBoard visualization (discussed in Section 9.2.5).

9.3.2   TENSOR LIBRARIES

Tensor expressions are implemented in the ATen C++ tensor library and some are still in the older tensor libraries (TH, THC, THNN, THCNN) from the Torch7 days. ATen implements many tensor types and different operators.

For the CPU, ATen and C2 use oneDNN in the shipped binaries with a fallback to Eigen and to OpenBLAS (particularly for PPC64). For the GPU, ATen uses cuBLAS, cuSolver, and MAGMA.

The ATen and C2 tensor libraries used in Caffe2 merged into the C10 library, which contain the tensor and storage data structures.

9.3.3   OTHER LIBRARIES

Other PyTorch-based libraries are torchvision, torchtext (used by PyText for Facebook's Messenger), and torchaudio. These libraries provide access to datasets and models in the vision, text, and audio domain, respectively.

fast.ai introduced the fastai high-level library that runs on top of PyTorch. It provides prebuilt models and additional tools, such as the LR finder. The library also runs on top of Swift4TF, but that project is less matured.

9.4   TVM

TVM is an Apache incubator project, and an end-to-end DL compiler stack for automatic code-generation across various hardware targets [CMJ+18]. TVM was developed by Tianqi Chen et al. at the University of Washington (UW). The project has several contributors from UW, Amazon Web Services (AWS), Qualcomm, Facebook, Google, Huawei, AMD, Microsoft, Cornell University, and University of California, Berkeley [Tvm19].

The TVM stack has two main levels of abstraction: a graph compiler and an operator-level compiler. TVM takes as input a model from MXNet, PyTorch/TorchScript, TensorFlow, Keras, CoreML, ONNX, and DarkNet and compiles it to the Relay IR (also known as NNVM v2) [Tvm19]. TVM is tightly integrated with MXNet with modules shared between the projects; both projects started at UW as part of the Deep Machine Learning Community (DMLC). The Relay IR is a statically-typed, complete (purely functional), modular, and extensible programming language. Relay provides common DL primitives, auto-differentiation, and mathematical optimizers.

TVM performs high-level graph optimization, on the Relay IR and then compiles into a low-level specification language called a *tensor expression* (TE). This language declaratively specifies the tensor operands, their shapes, and the operators, but the execution details are unspecified; thus, TVM decouples the definition of the expression with the execution. TVM borrows this decoupling idea from the Halide programming language [CMJ+18].

TVM defines a space of functionally-equivalent schedules for a TE and a given target. The space of schedules includes various loop transformations, cache localities, and vectorization strategies; a TE potentially has billions of schedules from all the possible combinations. A matrix multiplication TE can result in schedules with vanilla loops (see Algorithm 2.1), tiled loops, and accelerator intrinsics. Improving the constraints on the space of schedules is an important research area.

TVM borrows scheduling algorithms from Halide for CPUs and incorporates new algorithms for GPUs and accelerators. For a GPU and TPU-like accelerator, the space of schedules includes various strategies for thread cooperation and shared memory across the compute units. The space of schedules is usually the largest for a TPU-like accelerator. It includes hardware intrinsics for high-dimension tensor expressions and a hierarchical memory system with memory buffers and instructions for memory access. TVM uses a description of the hardware interface to narrow the scheduling space.

A goal of TVM is to automatically search over this space to obtain an efficient program configuration for a TE for a particular hardware target. One naive approach is to randomly sample the scheduling space, test each schedule on the target hardware, and return the sampled program configuration with the minimum runtime. Instead, TVM uses a simulated annealing algorithm to search the space of schedules, and *AutoTVM*, an ML-based performance predictor, to predict the runtime of a schedule without executing the schedule on the actual hardware.

AutoTVM learns a model that predicts the runtime of a schedule using an XGBoost algorithm, which is a computationally inexpensive ML algorithm. AutoTVM can be orders of magnitude faster than actual hardware runtime measurements [CG16]. Thus, this allows evaluating orders of magnitude more schedules and discovering a better one. Learning this model requires collecting training data using a dataset of schedules and measured runtime pairs. Transfer learning techniques can be used with new hardware or new TEs to reduce the required amount of training data.

The selected schedules are compiled using LLVM for CPUs, CUDA, OpenCL, or Metal for GPUs, or another back-end compiler for an accelerator. The compiled code is placed in a library with function pointers, and a higher-level program allocates input and output buffers and calls these functions during execution. TVM supports various deployment languages, including C++, Python, and Java.

The versatile tensor accelerator (VTA) is an open-source accelerator with an open-source microarchitecture and a software stack tightly integrated with TVM that can be prototyped on an FPGA or simulated on a laptop. Thus, VTA can facilitate the experimentation of custom optimizations across various back-end targets.

9.5   PLAIDML

PlaidML is an open-source (as of Aug. 2017) compiler stack developed and maintained by then vertex.ai and, as of Aug. 2018, part of Intel. PlaidML consumes a high-level static graph, such as ONNX, or others, and generates optimized code for various back-end targets. The most matured targets are GPUs and Movidius VPUs.

The PlaidML framework automatically generates efficient primitives from polyhedral tensor expressions, transforming graph-level operations requested by the graph compiler into optimized device-specific implementations. PlaidML compiles a high-level IR into target-dependent code: The high-level IR is mapped to the Tile IR using the Tile language capable of describing DL expressions. Like TVM's tensor expression, the Tile language is a differentiable DSL that represents mathematical formulas for the tensor expressions, and it is hardware agnostic.

A general polyhedral model allows for complex data dependencies. However, in a Tile contraction (a reduction operator that merges values across one or more indices), the only data dependency is in the aggregation. Tile only uses commutative and associative aggregation operations, so this dependency is only mildly restrictive. This narrow focus allows Tile's optimization to be more useful than general-purpose polyhedral optimizers.

The Tile IR lowers to a hardware-agnostic Stripe IR [ZB19]. The Stripe IR is then compiled via a series of hardware targeted optimizations and lowered to a hardware abstraction layer, accelerator runtime, or other hardware-appropriate code.

The Stripe IR uses hardware descriptions to constrain the optimization space using an affine tensor space. Stripe determines the optimal loop tiling and other loop permutations to reuse data across the memory hierarchy for a specific back-end target. The loop tiling parameters are selected based on hardware descriptors and adjusted via profile-guided optimizations. Stripe then produces an execution schedule for each primitive and inter-primitive data dependencies, including data movement instructions. PlaidML optimizations are also incorporated as an MLIR dialect.

9.6   GLOW

Glow (an abbreviation for **G**raph-**low**ering) is a DL compiler stack used for inference and training (the inference stack is more mature). The Glow compiler project is maintained by Facebook with committed support from Intel, Cadence, Esperanto, Marvell, Qualcomm, Bitmain, STMicroelectronics, Synposys, and Ceva [Fac20].

Glow is designed to compile a high-level graph supporting many operators to a low-level graph supporting a small number of linear algebra operators [Fac18]. The compiler passes can be shared across the various hardware targets. A separate hardware back-end compiler then consumes the low-level IR and generates executable code.

Glow takes as input a model from PyTorch's TorchScript or constructed via the C++ interface and compiles it to a high-level IR graph. Target-independent optimizations, such as automatic-differentiation and quantization to 8-bit integer if required, are applied to this high-level graph. Note that Glow does not use a polyhedral model as this has a long compilation time, which is not acceptable for JIT.

Glow compiles the high-level IR to a low-level instruction-based address-only (operands are typed pointers to buffers) IR via two lowerings. The first lowering decomposes the graph operators into convolution nodes and linear algebra operator nodes. For instance, a fully connected layer is transformed into a matrix multiplication node followed by a broadcasted add node (for the bias). Additional optimization passes occur on this mid-level IR. This graph is not SSA and is organized as a sequence of nodes with no control-flow.

The second lowering transforms the linear algebra nodes into a low-level instruction-based, address-only strongly-typed IR, known as IRGen. These instructions operate on tensors and are referenced by a hardware-independent address. The IRGen compiler passes determine the required memory allocation for these tensors and the possible in-place computations. The goal of this low-level IR is to facilitate optimizations by the back-end compiler.

The back-end compiler can consume either the mid-level or low-level IR (IRGen). It performs tensorization and code-generation for the specific hardware target. The back-end compiler may implement additional IRs with control-flow for low-level IR instructions, such as convolution.

Glow provides a CPU reference implementation to verify an accelerator's correct functionality. For CPU, Glow uses the LLVM compiler to optimize and generate code. The low-level IR can be AOT compiled (since the shapes and types of all the tensors are known) into machine code object files. These files are linked to some application with no further dependence on Glow (this is important for environments with limited memory, such as mobile devices). Alternatively, the low-level IR can execute code in JIT mode using a library of precompiled LLVM bitcode linear algebra micro-kernels written in C called `libjit`.

9.7   XLA

The Accelerated Linear Algebra (XLA) is a graph compiler developed and maintained by Google. XLA is used with TPUs, CPUs, and GPUs, and can be extended to other back-end targets. XLA is tightly integrated with TensorFlow and also supports PyTorch/Trace and Julia.

The TensorFlow APIs let the programmer explicitly invoke the XLA compiler on a subset of the TF graph (or the entire graph, if possible). The `tf2xla` compiler maps the TensorFlow subgraphs to the XLA High-Level Optimizer (HLO) IR. XLA decomposes the XLA HLO ops into basic functions, including element-wise ops, specialized NN ops (such as convolution), data layout reshape ops, control-flow ops, and data transfer ops [Goo20g]. Then, XLA fuses ops to reduce memory access overhead [Goo20c]. This optimized HLO IR maps to a back-end compiler for target-dependent optimizations and code-generation. XLA uses the LLVM compiler for code-generation on CPUs and GPUs, and a TPU compiler for TPUs. While XLA is a JIT compiler, it also provides AOT executable codegen compilation for some back-end tagets, such as CPUs.

In practice, XLA works well for a defined set of primitives, but supporting custom primitives can be a challenge [SL19]. This limits the adoption of XLA in the research community, where experimentation with new operators is common. In addition, XLA cannot compile tensors with dynamic shapes [BCD+18].

9.8   MLIR

One effort to improve the TensorFlow infrastructure and reduce the duplication of optimizations is the Multi-Level IR (MLIR). It was released in April 2019 by Google as a TensorFlow project, and later adopted as an LLVM project. While the initial front-end framework is TensorFlow, other frameworks can use it.

MLIR is a flexible ML SSA-based, typed-language, multilevel IR compiler infrastructure. MLIR is not a compiler but a compiler infrastructure; standard optimizations can be shared across the various levels of abstractions. It borrows many ideas from LLVM IR, both designed by Chris Lattner and other contributors, and has a library of optimization and compiler utilities. It has a flexible type system and supports dynamic tensor shapes and ranks. MLIR enables optimizations across various levels of abstractions from high-level optimizations with better control-flow representation to low-level compilers and executors that generate target machine code. The MLIR structure resembles the LLVM structure with modules, functions, blocks, and operations (note that in LLVM parlance, these are called instructions rather than operations, and in TVM parlance are called expressions). MLIR operators are the basic unit of MLIR code. Unlike LLVM, in MLIR the optimization passes are implicitly multithreaded.

MLIR IRs are called dialects. A dialect has a defined set of operations with input and output types and can express different levels of abstraction. Examples of dialects are the TensorFlow IR, XLA HLO, TFLite, Affine, and LLVM IR, and exclusively for GPUs: NVVM, SPIR-V, and ROCm. An affine dialect is a simplified polyhedral model with `for` loops and `if` control structure ops [Llv20]. A dialect provides invariants on the operators and a canonical representation. This canonicalization simplifies pattern-matching, verification, rewriting, and conversion to other dialects. Optimizations can be shared across dialects. Also, MLIR allows custom operators for a particular dialect.

Expressions can be written at multiple levels of abstraction. The high-level graph optimizations can use the TF dialect. The tensor optimizations (such as matrix multiplications and fusion) can use the XLA dialect, and the LLVM code-generation can use the LLVM dialect on supported hardware, all with the same infrastructure.

TensorFlow is gradually porting graph transformations to MLIR and unifying the interfaces to the back-end code generators [LS19]. Other hardware libraries or hardware vendor IRs can consume the MLIR and generate code for their respective back-end targets.

9.9   OTHERS

Other notable compilers include the following:

Halide was developed as a DSL for image processing [RBA+13]. Key Halide concepts can extend to DL compilers. TVM borrows many ideas from Halide, including decoupling the tensor expression from the schedule and defining the scheduling space.

Diesel was developed by Nvidia to generate efficient code for GPUs [ERR+18]. Diesels maps a DSL to a high-level graph and then lowers the graph to a Polyhedral IR. Optimization passes are applied to tile a loop for efficient parallelism between threads, warps, blocks, and SM. Diesel then generates CUDA code for various Nvidia GPU back-end architectures.

nGraph is an open-source C++ library for high-level compilation designed by Intel but no longer actively maintained. nGraph consumes a TensorFlow or ONNX computation graph, maps the subgraphs supported by nGraph to an nGraph IR (for TF models, the TF runtime handles nonsupported nodes), and performs high-level optimization passes, as shown in Figure 9.3 [SPE19].

Tensor Comprehension (TC) was developed by Facebook AI Lab and released in early 2018 [VZT+18]. Facebook appears to be prioritizing the Glow graph compiler. TC defines a scheduling space for GPUs using polyhedral methods and uses a JIT compiler to search for an efficient schedule. TC does not use ML to facilitate the selection of a schedule.

Tensor Algebra Compiler (taco) generates sparse tensor operators on a CPU [KKC+17].
DLVM has full control-flow and can be used for graph-level optimization [WSA18].
WELD is a DSL for data processing.

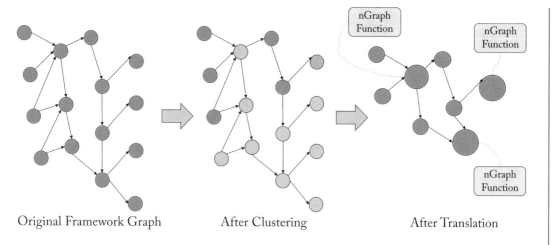

Figure 9.3: Graph-level optimizations used by nGraph (and typical in DL compilers). Various nodes are fused to reduce memory access overhead. Based on [SPE19].

In this chapter, we reviewed the importance of DL compilers to support the execution of models across diverse hardware targets. We detailed the DL compilers and software libraries used by hyperscalers and hardware vendors. The most popular frameworks (with built-in compilers) are TensorFlow and PyTorch, and the most popular compilers are TVM and XLA, with MLIR providing a compiler infrastructure. In the next chapter, we provide concluding remarks and discuss some of the future challenges and opportunities to advance DL.

CHAPTER 10

Opportunities and Challenges

In this concluding chapter, we discuss some of the opportunities and challenges ahead. The opportunities include using ML techniques to improve various aspects of the overall DL system. The challenges include security, interpretability, and the potential negative social impact, such as polarization, unemployment, power consumption, and copyright violations. We then provide some concluding remarks.

10.1 MACHINE LEARNING FOR DL SYSTEMS

ML (in particular, RL) can improve DL systems across various levels of the system stack. While more advances are needed to broadly use ML for DL system design, the success of MuZero finding good solutions in a huge search space suggests that similar techniques can provide solutions to other domains with large search spaces. These domains include integrated circuit designs, graph and primitive compilations, efficient and accurate topology designs, server and cluster configurations, and data center operations [YSE+20, SAH+20]. While ML for DL systems is a promising area, most of the work is in early stages, with limited adoption in production. Some areas where ML has been useful are [ZL17, HLL+19, RZQ+19, CMJ+18, MYP+19, KBC+18, WWS+19, MGP+18, WYL+20, EG16]:

- Integrated circuit (IC) design, which currently relies heavily on a human expert's experience and intuition.

- AutoML and NAS, discussed in more detail below.

- Transfer learning via Meta-learning.

- Schedule space search.

- Weight initialization.

- Layer quantization based on latency, energy, and accuracy requirement.

- Model compression.

- Index data structures (faster and with less memory than B-Trees and Bloom filters).

- Device placement for model parallelism.

- Power reduction in hyperscale data centers.

AutoML is the field of using ML to learn a DL model, tune the hyperparameters of an existing model, or compress a model. AutoML can reduce the data scientist's burden in model searching or parameter tuning at the expense of significant additional computations. **Neural architecture search** (NAS) is an area within AutoML that learns a model for a particular task, dataset, and hardware target by searching a constraint space of models or a subgraph within a larger graph. In practice, transfer learning is a more common technique used in production. In the future, hyperscalers may use NAS-based techniques to learn good models for a particular hardware architecture. Smaller companies can then use transfer learning on those learned models or use meta-learning methods, discussed later in this section.

Prominent NAS models are as follows: NASNet uses RL to learn an architectural building block using a smaller dataset (CIFAR-10) and then transfers that learned building block to learn a model using a more complex dataset (ImageNet) [ZL17]. NASNet has a superior Pareto curve on accuracy vs. the number of operations compared to human-designed models. AmoebaNet uses evolutionary algorithms to search the space faster than RL [RAH+19]. EfficientNet and EfficientDet are a family of models that uses a base network that scales across depth and width with better Pareto curve than NASNet and AmoebaNet [TL19, TPL19].

Differentiable architecture search (DARTS) assumes a continuous (differentiable) space of solutions, which enables the use of gradient-based optimizers. Adding regularization improves generalization [LSY19, ZES+20]. ProxylessNAS and FBNet use differentiable NAS applied to ConvNets to simultaneously learn the weights and the model for mobile phones deployment that are faster, smaller, and equally or more accurate than MobileNetV2 [CZH19, WDZ+19].

NAS-based techniques can reduce the number of required computations for a given topology. Efficient Neural Architecture Search (ENAS) uses Policy Gradient to learn a subgraph within a larger predefined graph [PGZ+18].

Meta-learning, also known as *learning to learn* and *few-shot learning*, is a form of transfer learning that learns a model from a few data samples by transferring knowledge from past learning experiences. The motivation is that knowledge learned from one task should benefit a different task; this is how humans learn.

There are two common approaches to meta-learning: metric-based where data samples are compared in a learned metric space (similar to nearest-neighbor classifiers) [VBL+17, SSZ17, LMR+19] and gradient-based where the model uses an update rule dictated by a meta-learner [MY17, FAL17, RRS+19, AES19]. A third approach combines these two approaches [FRP+20].

10.2   DEMOCRATIZING DL PLATFORMS

Several companies and cloud service providers developed higher-level platforms on top of the popular frameworks to facilitate a model's life cycle: data preparation, topology exploration,

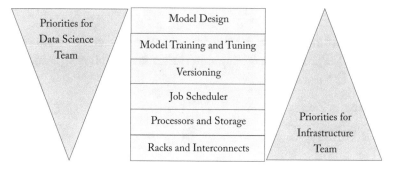

Figure 10.1: The data science and infrastructure teams have different priorities. Based on [BCC+19].

experiment tracking, model packaging, and model deployment at scale. At-scale deployment often uses Kubernetes (k8s) clusters or Spark clusters. These platforms provide a collaborative and secure environment and access to the latest ML libraries. These platforms are designed to meet the needs of the data scientists and the infrastructure teams, which typically have different priorities, as illustrated in Figure 10.1. Some of platforms are open-sourced. In the remainder of this section, we mention existing platforms that companies can adopt or emulate.

Platforms used for first-party users (that is, internal company users as opposed to third-party users, such as the external customers of cloud service providers) are as follows [HBB+18, Goo20e, Mic20, AAB+19, KR19, HDB17, HM19, Eid18, Met19, Met19b]:

- Facebook FBLearner

- Google TF Extended (TFX)

- Microsoft ML.NET

- eBay Krylov

- Uber Michelangelo

- AWS Eider

- Netflix Metaflow (integrated into AWS)

Platforms provided by cloud service providers for third-party users are as follows [Ama20, Goo20d, Mic20b, Ali20]:

- Amazon Sagemaker

- Google Cloud AI Platform

- Microsoft Azure cognitive services

- Alibaba PAI

Some of the above platforms can be deployed on-premise to facilitate switching between on-premise and on-cloud. Platforms targeting enterprises are as follows [Mlf20, Cor20, Nvi20, Int20c, Gui20, Ber19]:

- Intel Analytics Zoo

- Nvidia RAPIDS

- Databricks MLflow (to create models)

- Cortex (to deploy models at scale)

- Guild AI

- UC Berkeley RISE Ray

Some platforms facilitate the development and training of new models or the consumption of industry pre-trained models. As DL becomes widely adopted across industries, these platforms may become more critical.

10.3   SECURITY

Security expands all parts for the DL system stack from hardware to model robustness to data privacy. Attacks are increasing in scale and sophistication. In this section, we discuss two areas of active research: (1) adversarial ML and (2) data and model privacy. Although not discussed in further detail, DL is also used to improve security in domains, such as fraud detection, malware detection, vulnerability detection, and software verification [XLF+18, HDS+19].

Adversarial machine learning is the study of learning and preventing attacks. Adversarial attacks use tuned signals designed to deceive the model into producing a different than expected output. To illustrate, a correctly classified bus image can be imperceptibly perturbed to deceive a model to label it as an ostrich [SZS+14]. Adversarial attacks put in jeopardy applications where safety or security is critical, such as autonomous driving and biometric authentication.

Compressing a model makes it more vulnerable to these attacks by enlarging the magnitude of the adversarial noise [GWY+19, LGH19]. Training models robust to adversarial attacks can require larger models to converge to flatter minima (see Section 4.1), which in turn may require more computational resources [TSE+19].

There are two types of adversarial attacks: white-box and black-box attacks. In white-box attacks, the attacker knows the details of the target models, and in black-box attacks, the attacker does not have these details. Several techniques have been developed (none of them bulletproof) to increase robustness to adversarial attacks, including the following [ACW18, PMW+16, XEQ17, MC17, TKP+18, DAL+18, MMS+19, Nak19, LGH19, BV20, XZZ20]:

- defensive distillation to reduce the amplitude of the gradients, known as *gradient masking*, and smooth the model;

- reducing the bits per pixels in the input image and using spatial smoothing;

- training a model to modify adversarial examples, so they are correctly classified;

- augmenting the training dataset with adversarial examples;

- using models with larger capacity (more weights) than needed;

- optimizing robustness at smaller numerical representations;

- iteratively training a model with an adversary; and

- using the k-winners-take-all activation function.

Generative attacks use generative models to generate realistic data samples. These samples can deceive an authentication system or a human into believing the data is real. Mor et al. provide optimal strategies for the attacker and the authenticator systems and provide insights to design models robust to attacks [MPG+20].

Privacy is an area of active research. Key areas focused on preserving privacy are federated learning, GAN cryptography, homomorphic encryption, secured multiparty computations, and differential privacy.

Federated learning, discussed in Section 5.3, ensures that data stays local and is not transmitted to a centralized location. Training happens locally, and only the model updates are transmitted. However, some information about the local training data can be extracted from local updates [HAP17]. The updates should be encrypted before transmission and unencrypted only after the centralized location receives multiple models to preserve privacy [BIK+17].

GAN cryptography can facilitate training models that perform encryption and decryption [ACG+16]. Intel is developing homomorphic encryption tools to facilitate building models that operate on encrypted data. Homomorphic encryption methods, in theory, enable training and serving models using encrypted data; in practice, they require enormously more computations [Gen09]. Another more computationally feasible method is secure multiparty computations (SMPC), where parties jointly compute functions without revealing their inputs and outputs [ZZZ+19].

Differential privacy is an area of active research to train models without compromising the privacy of the training dataset [AA16, JYv19, LAG+19, DJS20, Goo20b, WZL+19]. Large models can memorize training data, and attackers may be able to extract information from a trained model. To illustrate, using a sentence completion tool an attacker types "The bank account of Amy Jones is", and the tool may regurgitate the actual account number if it is in the training dataset. To mitigate this vulnerability, Apple uses differential privacy technology adding some noise to the data in a user's device before such data is transmitted to Apple.

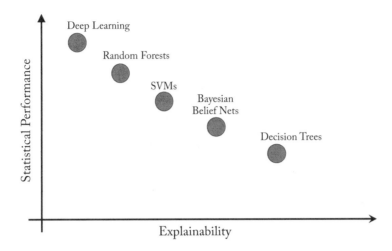

Figure 10.2: Algorithms that are more interpretable typically have lower accuracy. Note this is not shown to scale, but rather is a generalization of the algorithms' interpretability. Based on [Gun17].

10.4   INTERPRETABILITY

Interpretability is an area of active research to explain the reasons for the decisions, biases, and limitations of a given model. Limited interpretability is a barrier for some industries adopting DL algorithms despite their higher statistical performance. For instance, online credit applications should provide the reasons that a given loan was accepted or rejected. This *right-to-explanation* is required in some legal systems.

Interpretability methods can be applied to a topology using attention. Attention-based models learn to focus on the relevant inputs to produce a given output, which results in superior statistical performance while simultaneously provides interpretable insights [AP19, KZK+19, SLA+19].

BNN combine the strength of NNs and Bayesian models to estimate the uncertainty of a NN prediction [Nea95]. They can estimate uncertainty and provide performance guarantees. However, they are computationally expensive and require a good prior approximation to make them useful. BNNs are an active field of research.

An already trained model may be interpreted using activations, a saliency map, and testing concept vectors as follows: visualizing the activation features can provide insights into what a neuron or group of neurons learned but provides no insights into why a decision was made [ZF13, OSJ+18].

Another approach is using saliency maps to measure the impact of each input x_i in the output $p(z) : \frac{\partial p(z)}{\partial x_i}$. Salient maps are used in various domains, including in RL to gain insights

on the behavior of learned agents [GPV+20]. However, saliency map methods may lack reliability [AGM+18, HEK+19].

Google developed testing concept activation vectors (TCAV) to quantify the importance
of user-defined concepts in a model's output [KWG+18, Goo20f]. TCAV learns concepts from
examples. For instance, to determine the importance of stripes in classifying an image as a zebra,
a concept is learned using images of stripes, and then TCAV can test the model using this learned
concept. A current limitation is that the user needs to determine which concepts to test and needs
training samples to learn the concept.

Another aspect of interpretability is giving users information about the training of the
model. This information includes the objective function (what the model is mathematically designed to do), and the type of training data [MWZ+19, GMV+20]. Model developers should
explain where the model works and where it fails and possible biases in the model. Google calls
this the model card. This level of transparency is vital to accelerate DL adoption and mitigate
misuse or unintended consequences. The Partnership on AI is one effort in this direction. University of Washington's LIME and Google's What If Tool provide tools to analyze a model to
assist in this effort.

10.5   SOCIETY IMPACT

Artificial intelligence is improving multiple facets of society. Advances in DL and technology are
connecting the world. The monetary cost to communicate with people from around the world
is small and continues to decrease. NLP algorithms are breaking language barriers; in the near
future, persons without a common language may communicate in real-time with neural speech-
to-speech interpreters. In some areas, however, DL has a negative impact, and society should
address them. In this section, we briefly discuss crucial areas: polarization, algorithmic fairness,
unemployment, power consumption, and copyright violations.

DL is contributing to the polarization of society through personalized content. Companies that provide social media, news feeds, online videos, and other similar services, may optimize a metric related to maximizing user engagement. The result is content that reinforces,
rather than challenges, personal views limiting exposure to diverse postulations. In the author's
opinion, this limited exposure is the biggest threat that DL imposes to society because it can reverse the progress toward a more empathetic society. It is unclear how to address this challenge,
given the monetary incentives to maximize user engagement, and the competition between various content providers.

Data and algorithmic bias is a substantial issue in several production DL systems with
detrimental effects [LDR+18]. Training data is often skewed toward a demographic and incorporates human biases [BCZ+16]. Model developers should strive to collect unbiased data, and
provide a model card (discussed in Section 10.4). For instance, all other factors being equal, a
loan application should produce the same output across gender and race.

An area in which DL usually break down is in behavior prediction, such as recidivism, job success, terrorist risk, and at-risk kids [Nar19]. Given the current limited interpretability and biases in datasets, these types of behavior prediction applications should be avoided.

The community is developing tools to improve fairness. In addition to the tools mentioned in Section 10.4 for interpretability, IBM's AI Fairness 360 Toolkit can examine, report, and mitigate discrimination and bias in DL models using dozens of fairness metrics.

Technology advances increase human productivity and reduce the amount of human labor required to produce a set of goods. On the one hand, this increased productivity can result in shorter work hours, higher pay, and cheaper goods. On the other hand, as the cost of automation rapidly decreases, the need for human labor in various sectors of the economy diminishes. The rapid advancements in automation may result in significant unemployment if not adequately addressed. Some potential solutions are universal income, shorter work hours at higher pay, or re-education programs.

DL models require significant power budgets for training and serving. As they become more prominent in society, their large power consumption can negatively contribute to climate change. Fortunately, there is work-in-progress to develop smaller and more efficient models with smaller power budgets. In addition, DL algorithms are being used to find methods to mitigate climate change [RDK+19].

DL algorithms can be used to generate music, poetry, paintings, and voices and images of real persons using generative models. Models can generate synthetic content, including fake videos of real people. It is not clear how to simultaneously protect the individual rights of those people and the freedom-of-speech rights of the producer of those videos. Another challenge is that the generative model's training datasets can potentially contain copyright material, and it is not clear how to protect the rights of the original authors. In the United States, copyright material can be used to train search algorithms (using discriminative, rather than generative models) [Wik20].

These challenges are also a motivation to expand the public's knowledge of DL so that society can collectively find better ways to address them.

10.6 CONCLUDING REMARKS

The adoption of DL systems is rapidly growing and improving many technologies and services. In this book, we covered the algorithms, hardware, and compilers used in large-scale DL production systems by hyperscalers. We analyze the building blocks or primitives of computation graphs or topologies used in commercial applications. We explained the different types of workloads common at hyperscalers, academia, and smaller sized companies and highlighted the importance and prevalence of recommender models. We provided guidelines to train and debug a model so that it generalizes to data outside the training dataset, recommended the Ranger, Adam, SGDM, and LARS optimizers to train a model, and explained how to compute the gradients via backpropagation as a set of multiplications and additions. We highlighted the benefits of finding a batch size that results in high hardware utilization and low time-to-train.

As training times are getting prohibitively long on a single node, and some model sizes exceed a node's memory capacity, distributed training is required and used across hyperscalers. We provided guidelines for distributed training and recommended using a hybrid approach for large models: data parallelism is used across groups of super-nodes, and model parallelism is used within each super-node with 4–8 nodes per super-node. We highlighted pipeline parallelism as a promising approach to improve model parallelism, which is particularly beneficial for hardware platforms with large SRAM attempting to eliminate most DRAM accesses. However, further advances are needed to mitigate stalled weight updates and gain broad adoption.

Memory bandwidth is a common bottleneck in training and serving models. We discussed various smaller numerical formats to reduce the size of the model, which alleviates this bottleneck and results in more FPU units per die area. We recommended that production training hardware support $bf16$, $fp16$, and $fp32$ compute, and inference hardware support $fp16$, $bf16$, $fp8$, $int8$, and some $fp32$ compute. We also highlighted the statistical performance penalty when using $int8$ for various workloads. Hessian-based methods can guide the selection of quantizable layers and mitigate this penalty.

The rapid topology and algorithmic innovation motivate the adoption of a flexible platform that supports a broad spectrum of workloads, including disaggregated CPU to accelerator ratio, a standard form factor module, and an industry-standard interconnect to scale out the architecture. We discussed some of the performance vs. ease-of-programmability tradeoffs across various hardware designs used for DL, including the challenges of software-managed memory and the complexities of extracting high performance.

Central to DL systems are compilers to effectively lower a program to the target hardware. We detailed various optimization passes and highlighted operation fusion and loop tiling as pivotal optimizations to reduce memory accesses and efficiently use local memory. We discussed features in the popular frameworks TensorFlow and PyTorch, as well as the compilers TVM and XLA and others. Advancing compilers are critical to facilitate entry and competitiveness in the ecosystem.

We are in the early days of DL systems with exciting avenues to pursue and challenges to overcome. One of the most promising opportunities is using ML algorithms to improve the DL system stack. We reviewed potential ML for DL usages and highlighted some of the challenges the community needs to tackle.

I hope the concepts you learned in this book help you better engage with data scientists, hardware engineers, and compiler and performance engineers to tackle the compute, bandwidth, and memory demands, address the societal challenges, use ML to improve DL systems, and contribute to this exciting field. Thank you for your interest in DL systems and for the time you committed to studying this book.

Bibliography

[ABC+16] M. Abadi, P. Barham, J. Chen, et al. TensorFlow: a system for large-scale machine learning. *OSDI*, 2016. 184

[ACG+16] M. Abadi, A. Chu, I. Goodfellow, H. McMahan, I. Mironov, K. Talwar, and L. Zhang. Deep learning with differential privacy. *CCS*, Oct. 2016. 203

[AA16] M. Abadi and D. Andersen. Learning to protect communications with adversarial neural cryptography. Oct. 2016. 203

[AGM+18] J. Adebayo, J. Gilmer, M. Muelly, I. Goodfellow, M. Hardt, and B. Kim. Sanity checks for saliency maps. *NeurIPS*, Dec. 2018. 205

[ARS+20] D. Abts, J. Ross, J. Sparling, et al. Think fast: a tensor streaming processor (TSP) for accelerating deep learning workloads. *ISCA*, Jun. 2020. 154

[AMP+19] A. Agrawal, A. Modi, A. Passos, et al. TensorFlow Eager: a multi-stage, Python-embedded DSL for machine learning (slides). *MLSys*, Feb. 2019. 186, 187

[AAB+19] Z. Ahmed, S. Amizadeh, M. Bilenko, et al. Machine learning at Microsoft with ML .NET. *SIGKDD*, Jul. 2019. 201

[ALV08] M. Al-Fares, A. Loukissas, and A. Vahdat. A scalable, commodity data center network architecture. *SIGCOMM*, Oct. 2008. 150

[Ali20] Alibaba. Machine Learning Platform for AI. 2020. 201

[Ala18] J. Alammar. The illustrated transformer. June 2018. 64

[AHJ+18] D. Alistarh, T. Hoefler, M. Johansson, S. Khirirat, N. Konstantinov, and C. Renggli. The convergence of sparsified gradient methods. *NeurIPS*, Dec. 2018. 102

[AVG+15] L. Alvarez, L. Vilanova, M. Gonzalez, X. Martorell, N. Navarro, and E. Ayguade. Hardware-software coherence protocol for the coexistence of caches and local memories. *TC*, Jan. 2015. 134

[Ama19] Amazon. EC2 Inf1 Instances. 2019. 153

[Ama19b] Amazon. AWS re:Invent 2019: deliver high performance ML inference with AWS Inferentia. Dec. 2019. 153

[Ama20] Amazon. SageMaker. 2020. 201

[Amd67] G. Amdahl. Validity of the single processor approach to achieving large scale computing capabilities. *AFIPS*, Apr. 1967. 134

[Amd19] Amd. EPYC 7742. 2019. 152

[AAB+15] D. Amodei, R. Anubhai, E. Battenberg, et al. Deep Speech 2: end–to–end speech recognition in English and Mandarin. *ICML*, Dec. 2015. 66

[AC16] D. Amodei and J. Clark. Faulty reward functions in the wild. *OpenAI*, Dec. 2016. 70

[DH18] A. Dario and D. Hernandez. AI and compute. *OpenAI*, May 2018. 1, 99

[AES19] A. Antoniou, H. Edwards, and A. Storkey. How to train your MAML. *ICLR*, Mar. 2019. 200

[AP19] S. Arik and T. Pfister. ProtoAttend: attention-based prototypical learning. Sep. 2019. 204

[ABF+19] N. Arivazhagan, A. Bapna, O. Firat, et al. Massively multilingual neural machine translation in the wild: findings and challenges. July 2019. 61, 62

[ACB17] M. Arjovsky, S. Chintala, and L. Bottou. Wasserstein GAN. Jan. 2017. 14

[ADC11] T. Ashby, P. Diaz, and M. Cintra. Software-based cache coherence with hardware-assisted selective self-invalidations using Bloom filters. *TC*, Apr. 2011. 134

[AFO18] S. Ashkiani, M. Farach-Colton, and J. Owens. A dynamic hash table for the GPU. *IPDPS*, May 2018. 41

[ACW18] A. Athalye, N. Carlini, and D. Wagner. Obfuscated gradients give a false sense of security: circumventing defenses to adversarial examples. *ICML*, Jul. 2018. 202

[BKH16] J. Ba, J. Kiros, and G. Hinton. Layer normalization. July 2016. 38

[BGJ+18] V. Bacoyannis, V. Glukhov, T. Jin, J. Kochems, and D. Song. Idiosyncrasies and challenges of data driven learning in electronic trading. *NeurIPS*, Dec. 2018. 70

[Bai20] Baidu. Kunlun. 2020. 153

[BKK18] S. Bai, J. Kolter, and V. Koltun. An empirical evaluation of generic convolutional and recurrent networks for sequence modeling. Mar. 2018. 63

[BKK19] S. Bai, J. Kolter, and V. Koltun. Deep equilibrium models. *NeurIPS*, Dec. 2019. 97

[BTV06] H. Bay, T. Tuytelaars, and L. Van Gool. SURF: speeded up robust features. *ECCV*, 2006. 48

[BES+19] P. Balaprakash, R. Egele, M. Salim, V. Vishwanath, F. Xia, T. Brettin, and R. Stevens. Scalable reinforcement learning based neural architecture search for cancer deep learning research. *SC*, Nov. 2019.

[BV20] M. Balunovic and M. Vechev. Adversarial training and provable defenses: bridging the gap. *ICLR*, Feb. 2020. 202

[BHR18] L. Barroso, U. Holze, and P. Ranganathan. The datacenter as a computer: designing warehouse-scale machines. *M&C*, Oct. 2018. 3

[BLK+19] F. Belletti, K. Lakshmanan, W. Krichene, et al. Scaling up collaborative filtering data sets through randomized fractal expansions. Apr. 2019.

[Ben12] Y. Bengio. Practical recommendations for gradient-based training of deep architectures. *NNs: Tricks of the Trade*, Sep. 2012. 80, 92

[BBC+19] C. Berner, G. Brockman, B. Chan, et al. Dota 2 with large scale deep reinforcement learning. Dec. 2019. 70

[BCC+19] D. Berg, R. Chirravuri, R. Cledat, S. Goyal, F. Hamad, and V. Tuulos. Open-sourcing Metaflow, a human-centric framework for data science. *Netflix Tech Blog*, Dec. 2019. 201

[Ber19] Berkeley. Ray. 2019. 202

[BDD+20] M. Binkowski, J. Donahue, S. Dieleman, et al. High fidelity speech synthesis with adversarial networks. *ICLR*, Apr. 2020. 68

[BHH20] P. Blanchard, D. Higham, and N. Higham Accurately computing the log-sum-exp and softmax functions. *J. Num. Analysis*, Aug. 2020. 29, 120

[BCK+15] C. Blundell, J. Cornebise, K. Kavukcuoglu, and D. Wierstra. Weight uncertainty in neural networks. *ICML*, July 2015. 15

[BCZ+16] T. Bolukbasi, K. Chang, J. Zou, V. Saligrama, and A. Kalai. Man is to computer programmer as woman is to homemaker? Debiasing word embeddings. *NeurIPS*, Dec. 2016. 205

[BIK+17] K. Bonawitz, V. Ivanov, B. Kreuter, et al. Practical secure aggregation for privacy-preserving machine learning. *CCS*, Oct. 2017. 105, 203

[BHR+08] U. Bondhugula, A. Hartono, J. Ramanujam, and P. Sadayappan. A practical automatic polyhedral parallelizer and locality optimizer. *SIGPLAN*, June 2008.

[BAC16] U. Bondhugula, A. Acharya, and A. Cohen. The Pluto+ algorithm: A practical approach for parallelization and locality optimization of affine loop nests. *TOPLAS*, Apr. 2016.

[BLB17] A. Botev, G. Lever, and D. Barber. Nesterov's accelerated gradient and momentum as approximations to regularised update descent. *IJCNN*, Jul. 2017. 85

[BCD+18] T. Boyd, Y. Cao, S. Das, T. Joerg, and J. Lebar. Pushing the limits of GPU performance with XLA. Nov. 2018. 195

[BGL+93] J. Bromley, I. Guyon, Y. LeCun, E. Sackinger, and R. Shah. Signature verification using a "Siamese" time delay neural network. *NeurIPS*, Dec. 1993. 58

[Bro19] Y. Brovman. Complementary item recommendations at eBay scale. Feb. 2019. 45

[BMR+20] T. Brown, B. Mann, N. Ryder, M. Subbiah, et al. Language models are few-shot learners. May 2020.

[BCN06] C. Bucila, R. Caruana, and A. Niculescu-Mizil. Model compression. *SIGKDD*, Aug. 2006. 123

[BEP+18] Y. Burda, H. Edwards, D. Pathak, A. Storkey, T. Darrell, and A. Efros. Large-scale study of curiosity-driven learning. Aug. 2018. 70

[CZH19] H. Cai, L. Zhu, and S. Han. ProxylessNAS: direct neural architecture search on target task and hardware. *ICLR*, Feb. 2019. 200

[CBG+20] L. Cambier, A. Bhiwandiwalla, T. Gong, O. H. Elibol, M. Nekuii, and H. Tang. Shifted and squeezed 8-bit floating point format for low-precision training of deep neural networks. *ICLR*, Jan. 2020. 117, 118, 121

[HSW+18] Z. Cao, G. Hidalgo, T. Simon, S. Wei, and Y. Sheikh. OpenPose: realtime multi-person 2D pose estimation using part affinity fields. *CVPR*, Dec. 2018. 59

[CLN+17] I. Caspi, G. Leibovich, G. Novik, and S. Endrawis. Reinforcement Learning Coach. Dec. 2017. 70

[CMG+18] P. Castro, S. Moitra, C. Gelada, S. Kumar, and M. Bellemare. Dopamine: a research framework for deep reinforcement learning. Dec. 2018. 70

[CJL+16] W. Chan, N. Jaitly, Q. Le, and O. Vinyals. Listen, attend and spell: a neural network for large vocabulary conversational speech recognition. *ICASSP*, 2016. 67

[CFL20] O. Chang, L. Flokas, and H. Lipson. Principled weight initialization for hypernetworks. *ICLR*, Feb. 2020. 78

[CCS+17] P. Chaudhari, A. Choromanska, S. Soatto, et al. Entropy-SGD: biasing gradient descent into wide valleys. *ICLR*, Mar. 2017.

[CBH+11] N. Chawla, K. Bowyer, L. Hall, and W. Kegelmeyer. SMOTE: synthetic minority over-sampling technique. *JAIR*, June 2011. 88

[CHM+19] Y. Chebotar, A. Handa, V. Makoviychuk, M. Macklin, J. Issac, N. Ratliff, and D. Fox. Closing the sim-to-real loop: adapting simulation randomization with real world experience. *ICRA*, May 2019. 70

[CXZ+16] T. Chen, B. Xu, C. Zhang, and C. Guestrin. Training deep nets with sublinear memory cost. Apr. 2016. 96

[CES16] Y. Chen, J. Emer, and V. Sze. Eyeriss: a spatial architecture for energy-efficient dataflow for convolutional neural networks. *ISCA*, June 2016. 148

[CG16] T. Chen and C. Guestrin. XGBoost: a scalable tree boosting system. *SIGKDD*, Aug. 2016. 193

[CPS+17] L. Chen, G. Papandreou, F. Schroff, and H. Adam. Rethinking Atrous convolution for semantic image segmentation. June 2017. 57, 58

[CES17] Y. Chen, J. Emer, and V. Sze. Using dataflow to optimize energy efficiency of deep neural network accelerators. *MICRO*, June 2017. 147, 148

[CMJ+18] T. Chen, T. Moreau, Z. Jiang, et al. TVM: an automated end-to-end optimizing compiler for deep learning. *OSDI*, 2018. 192, 199

[CYC19] C. Chen, C. Yang, and H. Cheng. Efficient and robust parallel DNN training through model parallelism on multi-GPU platform. Oct. 2019. 103

[CZZ+19] C. Chen, M. Zhang, M. Zhang, Y. Liu, Y. Li, and S. Ma. Social attentional memory network: modeling aspect- and friend-level differences in recommendation. *WSDM*, Jan. 2019. 39

[CZL+19] Q. Chen, H. Zhao, W. Li, P. Huang, and W. Ou. Behavior sequence transformer for e-commerce recommendation in Alibaba. *DLP-KDD*, Aug. 2019. 47

[CMF+20] B. Chen, T. Medini, J. Farwell, S. Gobriel, C. Tai, and A. Shrivastava. SLIDE : in defense of smart algorithms over hardware acceleration for large-scale deep learning systems. *MLSys*, Mar. 2020. 41

[CYE+19] Y. Chen, T. Yang, J. Emer, and V. Sze. Eyeriss v2: a flexible accelerator for emerging deep neural networks on mobile devices. *JETCAS*, June 2019. 157

[CKH+16] H. Cheng, L. Koc, J. Harmsen, et al. Wide and deep learning for recommender systems. *DLRS*, Sep. 2016. 44, 46

[CWV+14] S. Chetlur, C. Woolley, P. Vandermersch, J. Cohen, J. Tran, B. Catanzaro, and E. Shelhamer. cuDNN: efficient primitives for deep learning. Dec. 2014. 134

[CCK+17] Y. Choi, M. Choi, M. Kim, J. Ha, S. Kim, and J. Choo. StarGAN: unified generative adversarial networks for multi-domain image-to-image translation. *CVPR*, Nov. 2017. 59

[CWV+18] J. Choi, Z. Wang, S. Venkataramani, P. Chuang, V. Srinivasan, and K. Gopalakrishnan. PACT: parameterized clipping activation for quantized neural networks. July 2018. 118

[Cho16] F. Chollet. Xception: deep learning with depthwise separable convolutions. *CVPR*, Oct. 2016. 53

[CB18] N. Choma and J. BrunaY. Graph neural networks for neutrino classification. *Big Data Summit*, Feb. 2018.

[CGC+14] J. Chung, C. Gulcehre, K. Cho, and Y. Bengio. Empirical evaluation of gated recurrent neural networks on sequence modeling. Dec. 2014. 35

[CFO+18] E. Chung, J. Fowers, K. Ovtcharov, et al. Serving DNNs in real time at datacenter scale with project Brainwave. *MICRO*, Mar. 2018. 117

[CAL+16] O. Cicek, A. Abdulkadir, S. Lienkamp, T. Brox, and O. Ronneberger. 3D U-Net: learning dense volumetric segmentation from sparse annotation. *MICCAI*, June 2016. 57

[Cor20] Cortex. Deploy machine learning models in production. 2020. 202

[CAS16] P. Covington, J. Adams, and E. Sargin. Deep neural networks for YouTube recommendations. *RecSys*, Sep. 2016. 46

[DB19] W. Dai and D. Berleant. Benchmarking contemporary deep learning hardware and frameworks: a survey of qualitative metrics. *CogMI*, Dec. 2019. 157

[DAM+16] D. Das, S. Avancha, D. Mudigere, et al. Distributed deep learning using synchronous stochastic gradient descent. Feb. 2016. 33

[Dal16] B. Dally. High-performance hardware for machine learning. *ENN*, Feb. 2017. 129

[DMM+18] D. Das, N. Mellempudi, D. Mudigere, et al. Mixed precision training of convolutional neural networks using integer operations. *ICLR*, Feb. 2018. 116

[DPG+14] Y. Dauphin, R. Pascanu, C. Gulcehre, K. Cho, S. Ganguli, and Y. Bengio. Identifying and attacking the saddle point problem in high-dimensional non-convex optimization. *NeurIPS*, Dec. 2014. 80

[DKA+19] S. Dave, Y. Kim, S. Avancha, K. Lee, and A. Shrivastava. DMazeRunner: executing perfectly nested loops on dataflow accelerators. *TECS*, Oct. 2019. 148

[Daw20] DAWNBench. DAWNBench: an end-to-end deep learning benchmark and competition. 2020. 157

[DCJ19] M. Dacrema, P. Cremonesi, and D. Jannach. Are we really making much progress? A worrying analysis of recent neural recommendation approaches. *RecSys*, Sep. 2019. 44

[Dee19] DeepBench. Benchmarking deep learning operations on different hardware. 2019. 157

[DGY+74] R. Dennard, F. Gaensslen, H. Yu, V. Rideout, E. Bassous, and A. LeBlanc. Design of ion-implanted MOSFET's with very small physical dimensions. *JSSC*, Oct. 1974. 127, 131

[DAM+19] D. Dennis, D. Acar, V. Mandikal, V. Sadasivan, H. Simhadri, V. Saligrama, and P. Jain. Shallow RNNs: a method for accurate time-series classification on tiny devices. *NeurIPS*, Dec. 2019. 62

[Dev17] J. Devlin. Sharp models on dull hardware: fast and accurate neural machine translation decoding on the CPU. May 2017. 111

[DCL+18] J. Devlin, M. Chang, K. Lee, and K. Toutanova. BERT: pre-training of deep bidirectional transformers for language understanding. Oct. 2018. 63

[DAL+18] G. Dhillon, K. Azizzadenesheli, Z. Lipton, et al. Stochastic activation pruning for robust adversarial defense. *ICLR*, Mar. 2018. 202

[dDF+19] F. de Dinechin, L. Forget, J. Muller, and Y. Uguen. Posits: the good, the bad and the ugly. *CoNGA*, Mar. 2019. 118

[DSK+19] Y. Ding, J. Sohn, M. Kawczynski, et al. A deep learning model to predict a diagnosis of Alzheimer disease by using F-FDG PET of the brain. *Radiology*, Feb. 2019.

[DPB+17] L. Dinh, R. Pascanu, S. Bengio, and Y. Bengio. Sharp minima can generalize for deep nets. *ICML*, Aug. 2017. 77

[DWO+19] Z. Doctor, D. Wysocki, R. O'Shaughnessy, D. Holz, and B. Farr. Black hole coagulation: modeling hierarchical mergers in black hole populations. Nov. 2019. 5

[DDV+20] T. Domhan, M. Denkowski, D. Vilar, X. Niu, F. Hieber, and K. Heafield. The Sockeye 2 neural machine translation toolkit at AMTA 2020. Aug. 2020. 65

[Don19] L. Dong. eBay's hyperscale platforms. Sep. 2019. 44, 118

[DYC+19] Z. Dong, Z. Yao, Y. Cai, D. Arfeen, A. Gholami, M. Mahoney, and K. Keutzer. HAWQ-V2: Hessian aware trace-weighted quantization of neural networks. Nov. 2019. 76, 81, 120, 121

[Doz16] T. Dozat. Incorporating Nesterov momentum into Adam. *ICLR*, May 2016. 85

[DMM+19] N. Dryden, N. Maruyama, T. Moon, T. Benson, M. Snir, and B. Van Essen. Channel and filter parallelism for large-scale CNN training. *SC*, Nov. 2019. 102

[DJS20] M. Du, R. Jia, and D. Song. Robust anomaly detection and backdoor attack detection via differential privacy. *ICLR*, Feb. 2020. 203

[DHS11] J. Duchji, E. Hazan, and Y. Singer. Adaptive subgradient methods for online learning and stochastic optimization. *JMLR*, July 2011. 85

[Efr20] A. Efrati. AI startups proliferate as businesses look for savings. *The Information*, Aug. 2020.

[ERR+18] V. Elango, N. Rubin, M. Ravishankar, H. Sandanagobalane, and V. Grover. Diesel: DSL for linear algebra and neural net computations on GPUs. *MAPL*, June 2018. 196

[Eid18] Eider. Expo Demo. *NeurIPS*, Dec. 2018. 201

[ENG+18] A. Eisenman, M. Naumov, D. Gardner, M. Smelyanskiy, S. Pupyrev, K. Hazelwood, A. Cidon, and S. Katti. Bandana: using non-volatile memory for storing deep learning models. Nov. 2018. 47

[ETT15] T. Erez, Y. Tassa, and E. Todorov. Simulation tools for model-based robotics: comparison of Bullet, Havok, MuJoCo, ODE and PhysX. *ICRA*, May 2015. 70

[EBA+11] H. Esmaeilzadeh, E. Blem, R. S. Amant, K. Sankaralingam, and D. Burger. Dark silicon and the end of multicore scaling. *ISCA*, June 2011.

[EG16] R. Evans and J. Gao. DeepMind AI reduces Google data centre cooling bill by 40 percent. July 2016. 199

[Fac18] Facebook. Glow IR. Oct. 2018. 194

[Fac20] Facebook. Compiler for neural network hardware accelerators. Feb. 2020. 194

[FHY19] F. Farshchi, Q. Huang, and H. Yun. Integrating NVIDIA deep learning accelerator (NVDLA) with RISC-V SoC on FireSim. *EMC2*, Dec. 2019. 157

[Fel19] M. Feldman. AI recommendation systems get a GPU makeover. 2018. 45

[Fel19b] A. Feldman. Cerebras deploys the CS-1, the industry's fastest AI computer, at Argonne National Lab. Nov. 2019. 156

[FGM+10] P. Felzenszwalb, R. Girshick, D. McAllester, and D. Ramanan. Object detection with discriminatively trained part-based models. *PAMI*, Sep. 2010. 5

[Fey20] M. Fey. PyTorch geometric documentation. 2020. 47

[FL19] M. Fey and J. Lenssen. Fast graph representation learning with PyTorch geometric. Mar. 2019. 13, 47

[FAL17] C. Finn, P. Abbeel, and S. Levine. Model-agnostic meta-learning for fast adaptation of deep networks. *ICML*, July 2017. 200

[FWT11] V. Firoiu, W. Whitney, and J. Tenenbaum. Beating the world's best at Super Smash Bros. with deep reinforcement learning. May 2017. 70

[FRP+20] S. Flennerhag, A. Rusu, R. Pascanu, F. Visin, H. Yin, and R. Hadsell. Meta-learning with warped gradient descent. *ICLR*, Apr. 2020. 200

[FC19] J. Frankle and M. Carbin. The lottery ticket hypothesis: finding sparse, trainable neural networks. *ICLR*, Mar. 2019. 122

[FLP+99] M. Frigo, C. E. Leiserson, H. Prokop, and S. Ramachandran. Cache-oblivious algorithms. 1999. 172

[Gab46] D. Gabor. Theory of communication. Part 1: the analysis of information. *Radio & Comm. Eng.*, Nov. 1946. 48

[GZY+20] T. Gale, M. Zaharia, C. Young, and Erich Elsen. Sparse GPU kernels for deep learning. June 2020. 122

[GCL+19] J. Gauci, E. Conti, Y. Liang, et al. Horizon: Facebook's open source applied reinforcement learning platform. Sep. 2019. 70

[GMV+20] T. Gebru, J. Morgenstern, B. Vecchione, J. Vaughan, H. Wallach, H. Daume III, and K. Crawford. Datasheets for datasets. Mar, 2019. 205

[GAG+17] J. Gehring, M. Auli, D. Grangier, D. Yarats, and Y. Dauphin. Convolutional sequence to sequence learning. *ICML*, May 2017. 63

[GRM+18] R. Geirhos, P. Rubisch, C. Michaelis, M. Bethge, F. Wichmann, and W. Brendel. ImageNet-trained CNNs are biased towards texture; increasing shape bias improves accuracy and robustness. Nov. 2018. 49

[Gen09] C. Gentry. A fully homomorphic encryption scheme. Sep. 2009. 203

[GAB+18] E. Georganas, S. Avancha, K. Banerjee, D. Kalamkar, G. Henry, H. Pabst, and A. Heinecke. Anatomy Of high-performance deep learning convolutions on SIMD architectures. *SC*, Aug. 2018. 134

[GSC99] F. Gers, J. Schmidhuber, and F. Cummins. Learning to forget: continual prediction with LSTM. *ICANN*, Sep. 1999. 79

[Gha17] A. Gharakhanian. Generative adversarial networks–hot topic in machine learning. *KDnuggets*, Jan. 2017. 14

[GAJ+18] A. Gholami, A. Azad, P. Jin, K. Keutzer, and A. Buluc. Integrated model, batch, and domain parallelism in training neural networks. *SPAA*, July 2018. 102

[GLH+19] S. Ghose, T. Li, N. Hajinazar, D. Cali, and O. Mutlu. Understanding the interactions of workloads and DRAM types: a comprehensive experimental study. Oct. 2019. 137

[GCH+20] B. Ginsburg, P. Castonguay, O. Hrinchuk, et al. Stochastic gradient methods with layer-wise adaptive moments for training of deep networks. Feb. 2020. 85

[GBB11] X. Glorot, A. Bordes, and Y. Bengio. Deep sparse rectifier neural networks. *AISTATS*, 2011. 28

[GB10] X. Glorot and Y. Bengio. Understanding the difficulty of training deep feedforward neural networks. *AISTATS*, 2010. 78

[GPM+14] I. Goodfellow, J. Pouget-Abadie, M. Mirza, et al. Generative adversarial networks. Jun. 2014. 13

[Goo19] Google. MLIR: a new intermediate representation and compiler framework. Apr. 2019. 188

[Goo20] Google. Embeddings: translating to a lower-dimensional space. 2020. 39

[Goo20b] Google. C++ differential privacy library. Feb. 2020. 203

[Goo20c] Google. TensorFlow XLA index. Feb. 2020. 195

[Goo20d] Google. AI Platform. 2020. 201

[Goo20e] Google. TensorFlow Extended (TFX) is an end-to-end platform for deploying production ML pipelines. 2020. 201

[Goo20f] Google. TensorFlow TCAV. 2020. 205

[Goo20g] Google. TensorFlow-XLA Operation Semantics. 2020. 195

[Gvd08] K. Goto, and R. van de Geijn. Anatomy of high-performance matrix multiplication. *TOMS*, May 2008. 134

[Gra19] GraphCore. Microsoft and Graphcore collaborate to accelerate artificial intelligence. 2019. 154

[Gra20] Graphcore. Intelligent processing unit. July 2020. 156

[GSK+17] K. Greff, R. Srivastava, J. Koutník, B. Steunebrink, and J. Schmidhuber. LSTM: a search space dyssey. *TNNLS*, Oct. 2017. 35

[GW00] A. Griewank and A. Walther. Algorithm 799: revolve: an implementation of check-pointing for the reverse or adjoint mode of computational differentiation. *TOMS*, Mar. 2000. 96

[GMY+19] H. Guan, A. Malevich, J. Yang, Jongsoo Park, and H. Yuen. Post-training 4-bit quantization on embedding tables. *NeurIPS*, Dec. 2019. 117, 118

[GWY+19] S. Gui, H. Wang, C. Yu, H. Yang, Z. Wang, and J. Liu. Model compression with adversarial robustness: a unified optimization framework. *NeurIPS*, Dec. 2019. 123, 202

[Gui20] GuildAI. The ML Engineering Platform. 2020. 202

[Gun17] D. Gunning. Explainable Artificial Intelligence (XAI). *DARPA*, Nov. 2017. 204

[GPV+20] P. Gupta, N. Puri, S. Verma, D. Kayastha, S. Deshmukh, B. Krishnamurthy, and S. Singh. Explain your move: understanding agent actions using focused feature saliency. *ICLR*, 2020. 205

[GTY+17] H. Guo, R. Tang, Y. Ye, Z. Li, and X. He. DeepFM: a factorization-machine based neural network for CTR prediction. Mar. 2017. 46

[Gus17] J. Gustafson. Posit arithmetic. 2017. 118

[Hab19] Habana Labs. Goya inference platform white paper. Aug. 2019. 116

[Hab19b] Habana Labs. System-1. June 2019. 155

[HKK16] D. Han, J. Kim, and J. Kim. Deep pyramidal residual networks. *CVPR*, Oct. 2016. 123

[HPN+17] S. Han, J. Pool, S. Narang, et al. DSD: dense-sparse-dense training for deep neural networks. *ICLR*, Feb. 2017. 123

[HRM+19] A. Hard, K. Rao, R. Mathews, et al. Federated learning for mobile keyboard prediction. Feb. 2019. 103

[HNP+18] A. Harlap, D. Narayanan, A. Phanishayee, V. Seshadri, N. Devanur, G. Ganger, and P. Gibbons. PipeDream: fast and efficient pipeline parallel DNN training. June 2018. 102, 103

[Har18] F. Hartmann. Federated learning for Firefox. Aug. 2018. 103

[Has18] M. Hassan. AlexNet-1.png. 2018. 50

[Haz18] K. Hazelwood. Applied machine learning at Facebook: an infrastructure perspective. Sep. 2018. 104, 152

[HBB+18] K. Hazelwood, S. Bird, D. Brooks, et al. Applied machine learning at Facebook: a datacenter infrastructure perspective. *HPCA*, Feb. 2018. 201

[Haz20] K. Hazelwood. Deep learning: it's not all about recognizing cats and dogs. *SAIS*, June 2020. 3, 22

[HBG+08] H. He, Y. Bai, E. A. Garcia, and S. Li. ADASYN: adaptive synthetic sampling approach for imbalanced learning. *IJCNN*, June 2008. 88

[HZR+15] K. He, X. Zhang, S. Ren, and J. Sun. Deep residual learning for image recognition. *CVPR*, Dec. 2015. 53

[HZR+15] K. He, X. Zhang, S. Ren, and J. Sun. Delving deep into rectifiers: surpassing human-level performance on ImageNet classification. *ICCV*, Feb. 2015. 28, 78

[HZR+15] K. He, X. Zhang, S. Ren, and J. Sun. Spatial pyramid pooling in deep convolutional networks for visual recognition. Apr. 2015. 34

[HGD+17] K. He, G. Gkioxari, P. Dollar, and R. Girshick. Mask R-CNN. *ICCV*, Mar. 2017. 57, 122

[HLZ+17] X. He, L. Liao, H. Zhang, L. Nie, X. Hu, and T. Chua. Neural collaborative filtering. *ICIWWW*, Apr. 2017. 46, 47

[HSP+19] Y. He, T. Sainath, R. Prabhavalkar, et al. Streaming end-to-end speech recognition for mobile devices. *ICASSP*, Apr. 2019. 12, 67, 117

[HLL+19] Y. He, J. Lin, Z. Liu, H. Wang, L. Li, and S. Han. AMC: AutoML for model compression and acceleration on mobile devices. *ECCV*, Jan. 2019. 121, 199

[HAP+19] K. Hegde, H. Asghari-Moghaddam, M. Pellauer, et al. ExTensor: an accelerator for sparse tensor algebra. *MICRO*, Oct. 2019. 148

[HIB+19] P. Henderson, R. Islam, P. Bachman, J. Pineau, D. Precup, and D. Meger. Deep Reinforcement Learning that Matters. Jan. 2019. 71

[HG16] D. Hendrycks and K. Gimpel. Gaussian error linear units (GELUs). June 2016. 28

[HDB17] J. Hermann and M. Del Balso. Meet Michelangelo: Uber's machine learning platform. Sep. 2017. 201

[HMv+17] M. Hessel, J. Modayil, H. van Hasselt, et al. Rainbow: combining improvements in deep reinforcement learning. *AAAI*, Oct. 2017. 71

[HR15] T. Highlander and A. Rodriguez. Very efficient training of convolutional neural networks using fast Fourier transform and overlap-and-add. *BMVA*, Sep. 2015. 33

[HSS12] G. Hinton, N. Srivastava, and K. Swersky. RMSProp: divide the gradient by a running average of its recent magnitude. *Coursera*, 2012. 85

[HVD15] G. Hinton, O. Vinyals, and J. Dean. Distilling the knowledge in a neural network. Mar. 2015. 68, 123

[HAP17] B. Hitaj, G. Ateniese, and F. Perez-Cruz. Deep models under the GAN: information leakage from collaborative deep learning. *SIGSAC CCS*, Sep. 2017. 105, 203

[HS97] S. Hochreiter and J. Schmidhuber. Flat minima. *Neural Comp.*, Jan. 1997. 75

[HS97] S. Hochreiter and J. Schmidhuber. Long short-term memory. *Neural Comp.*, Nov. 1997. 79

[HHS17] E. Hoffer, I. Hubara, and D. Soudry. Train longer, generalize better: closing the generalization gap in large batch training of neural networks. *NeurIPS*, Dec. 2017. 78, 93

[HM19] A. Holler and M. Mui. Evolving Michelangelo model representation for flexibility at scale. Oct. 2019. 201

[HEK+19] S. Hooker, D. Erhan, P. Kindermans, and B. Kim. A benchmark for interpretability methods in deep neural networks. *NeurIPS*, Dec. 2019. 205

[HSW89] K. Hornik, M. Stinchcombe, and H. White. Multilayer feedforward networks are universal approximators. *NNs*, Mar. 1989. 9

[Hor14] M. Horowitz. 1.1 Computing's energy problem (and what we can do about it). *ISSCC*, Feb. 2014. 129, 138

[Hou19] J. Hou. New research on quantization could revolutionize power-efficient AI. July 2019. 153

[HZC+17] A. G. Howard, M. Zhu, B. Chen, et al. MobileNets: efficient convolutional neural networks for mobile vision applications. Apr. 2017. 54

[HSA+19] J. Hu, L. Shen, S. Albanie, G. Sun, and E. Wu. Squeeze-and-excitation networks. *CVPR*, May 2019. 59

[HLG+19] W. Hu, B. Liu, J. Gomes, M. Zitnik, P. Liang, V. Pande, and J. Leskovec. Strategies for pre-training graph neural networks. *ICLR*, Sep. 2019.

[HZS+19] W. Hua, Y. Zhou, C. Sa, Z. Zhang, and G. Suh. Channel gating neural networks. *NeurIPS*, Dec. 2019. 122

[HLv+16] G. Huang, Z. Liu, L. van der Maaten, and K. Weinberger. Densely connected convolutional networks. *CVPR*, Aug. 2016. 53

[HLP+17] X. Huang, Y. Li, O. Poursaeed, J. Hopcroft, and S. Belongie. Stacked generative adversarial networks. *CVPR*, June 2017. 59

[HCB+19] Y. Huang, Y. Cheng, A. Bapna, et al. GPipe: efficient training of giant neural networks using pipeline parallelism. *NeurIPS*, Dec. 2019. 103

[HDS+19] D. Huang, P. Dhariwal, D. Song, and I. Sutskever. GamePad: a learning environment for theorem proving. *ICLR*, 2019. 202

[Hua19] Huawei. Ascend 910 AI processor. 2019.

[Hug15] C. Hughes. Single-instruction multiple-data execution. *M&C*, May 2015.

[HS14] K. Hwang and W. Sung. Fixed-point feedforward deep neural network design using weights +1, 0, and -1. *SiPS*, Oct. 2014. 119

[IHM+16] F. Iandola, S. Han, M. Moskewicz, K. Ashraf, W. Dally, and K. Keutzer. SqueezeNet: AlexNet-level accuracy with 50x fewer parameters and <0.5MB model size. *CVPR*, Feb. 2016. 59

[Ibm20] IBM. IBM reveals next-generation IBM POWER10 processor. Aug. 2020. 152

[Int18] Intel. Knowledge Distillation. 2018. 124

[Int19] Intel. Next-generation Intel Xeon Scalable processors to deliver breakthrough platform performance with up to 56 processor cores. Aug. 2019. 152

[Int19b] Intel. Aurora SuperComputer. Nov. 2019. 153

[Int20] Intel. Innovation through intelligence. Jan. 2020. 152

[Int20b] Intel. Intel architecture instruction set extensions and future features programming reference. June 2020. 152

[Int20c] Intel. Analytics zoo. 2020. 202

[IS15] S. Ioffe and C. Szegedy. Batch normalization: accelerating deep network training by reducing internal covariate shift. Feb. 2015. 37, 51

[Iof17] S. Ioffe. Batch renormalization: towards reducing minibatch dependence in batch-normalized models. *NeurIPS*, Dec. 2017. 38, 81

[IZZ+16] P. Isola, J. Zhu, T. Zhou, and A. Efros. Image-to-image translation with conditional adversarial networks. *CVPR*, Nov. 2016. 59

[Iva71] A. Ivakhnenko. Polynomial theory of complex systems. *SMC*, Oct. 1971. 121

[IPG+19] P. Izmailov, D. Podoprikhin, T. Garipov, D. Vetrov, and A. Wilson. Averaging weights leads to wider optima and better generalization. *UAI*, Feb. 2019. 83, 85

[Jad19] A. Jadhav. Applications of graph neural networks. Feb. 2019. 13

[JJN+19] P. Jain, A. Jain, A. Nrusimha, A. Gholami, P. Abbeel, K. Keutzer, I. Stoica, and J. Gonzalez. Checkmate: breaking the memory wall with optimal tensor rematerialization. Oct. 2019. 96, 97

[JFZ+19] M. Janner, J. Fu, M. Zhang, and S. Levine. When to trust your model: model-based policy optimization. *NeurIPS*, Dec. 2019. 72

[JYS19] D. Jauk, D. Yang, and M. Schulz. Predicting faults in high performance computing systems: an in-depth survey of the state-of-the-practice. *SC*, Nov. 2019.

[Jax20] Jax. Composable transformations of Python+NumPy programs: differentiate, vectorize, JIT to GPU/TPU, and more. Feb. 2020. 182

[JJM+18] Y. Jia, M. Johnson, W. Macherey, et al. Leveraging weakly supervised data to improve end-to-end speech-to-text translation. *ICASSP*, Nov. 2018. 69

[JWB+19] Y. Jia, R. Weiss, F. Biadsy, W. Macherey, M. Johnson, Z. Chen, and Y. Wu. Direct speech-to-speech translation with a sequence-to-sequence model. Apr. 2019. 69

[JZW+18] Y. Jia, Y. Zhang, R. Weiss, et al. Transfer learning from speaker verification to multispeaker text-to-speech synthesis. *NeurIPS*, Dec. 2018. 68, 69

[JZA18] Z. Jia, M. Zaharia, and A. Aiken. Beyond data and model parallelism for deep neural networks. *ML*, July 2018. 102

[JHJ+20] Y. Jiao, L. Han, R. Jin, et al. 12nm programmable convolution-efficient neural-processing-unit chip achieving 825 TOPS. *ISSCC*, Feb. 2020. 118, 153

[JGK18] P. Jin, B. Ginsburg, and K. Keutzer. Spatially parallel convolution. *ICLR*, 2018. 102

[Joh18] J. Johnson. Rethinking floating point for deep learning. *NeurIPS*, Dec. 2018. 118

[JS18] M. Johnson and B. Stevens. Pruning hypothesis comes of age. *Nature*, Feb. 2018. 121

[JYv19] J. Jordon, J. Yoon, and M. van der Schaar. PATE-GAN: generating synthetic data with differential privacy guarantees. *ICLR*, Feb. 2019. 203

[JYP+17] N. Jouppi, C. Young, N. Patil, D. Patterson, et al. In-datacenter performance analysis of a tensor processing unit. *ISCA*, June 2017. 117

[JYK+20] N. Jouppi, D. Yoon, G. Kurian, S. Li, N. Patil, J. Laudon, C. Young, and D. Patterson. A domain-specific supercomputer for training deep neural networks. *CACM*, July 2020. 99, 153

[JZS15] R. Jozefowicz, W. Zaremba, and I. Sutskever. An empirical exploration of recurrent network architectures. *ICML*, July 2015. 79

[KZK+19] D. Kaji, J. Zech, J. Kim, S. Cho, N. Dangayach, A. Costa, and E. Oermann. An attention based deep learning model of clinical events in the intensive care unit. Feb. 2019. 204

[KES+18] N. Kalchbrenner, E. Elsen, K. Simonyan, et al. Efficient neural audio synthesis. June 2018. 68

[KMM+19] D. Kalamkar, D. Mudigere, N. Mellempudi, et al. A study of bfloat16 for deep learning training. June 2019. 115, 116

[KMH+20] J. Kaplan, S. McCandlish, T. Henighan, T. Brown, B. Chess, R. Child, S. Gray, A. Radford, J. Wu, and D. Amodei. Scaling laws for neural language models. Jan. 2020. 77

[KBA18] S. Karandikar, D. Biancolin, and A. Amid. FireSim. 2018. 157

[KCH+19] S. Karita, N. Chen, T. Hayashi, et al. A comparative study on transformer vs RNN in speech applications. Sep. 2019. 63

[Kar19] A. Karpathy. A recipe for training neural networks. Apr. 2019. 79, 91

[KLA19] T. Karras, S. Laine, and T. Aila. A style-based generator architecture for generative adversarial networks. *CVPR*, Mar. 2019. 59

[KR19] S. Katariya and A. Ramani. eBay's transformation to a modern AI platform. Dec. 2019. 201

[KMN+17] N. Keskar, D. Mudigere, J. Nocedal, M. Smelyanskiy, and P. Tang. On large-batch training for deep learning: generalization gap and sharp minima. *ICLR*, Apr. 2017. 76, 81

[KS17] N. Keskar and R. Socher. Improving generalization performance by switching from Adam to SGD. Dec. 2017. 84

[KDT+05] J. Kim, W. Dally, B. Towles, and A. Gupta. Microarchitecture of a high-radix router. *ISCA*, June 2005. 151

[KDS+08] J. Kim, W. Dally, S. Scott, and D. Abts. Technology-driven, highly-scalable drag-onfly topology. *ISCA*, June 2008. 151

[KWG+18] B. Kim, M. Wattenberg, J. Gilmer, C. Cai, J. Wexler, F. Viegas, and R. Sayres. Interpretability beyond feature attribution: quantitative testing with concept activation vectors (TCAV). *ICML*, June 2018. 205

[KKS+19] C. Kim, S. Kang, D. Shin, S. Choi, Y. Kim, and H. Yoo. A 2.1TFLOPS/W mobile deep RL accelerator with transposable PE array and experience compression. *ISSCC*, Feb. 2019. 70

[KB17] D. Kingma and J. Ba. Adam: a method for stochastic optimization. *ICLR*, Jan. 2017. 84

[KKC+17] F. Kjolstad, S. Kamil, S. Chou, D. Lugato, and S. Amarasinghe. The tensor algebra compiler. *OOPSLA*, Oct. 2017. 196

[KUM+17] G. Klambauer, T. Unterthiner, A. Mayr, and S. Hochreiter. Self-normalizing neural networks. *NeurIPS*, Dec. 2017. 28

[Kod19] R. Koduri. Intel unveils new GPU architecture with high-performance computing and AI acceleration, and oneAPI software stack with unified and scalable abstraction for heterogeneous architectures. *Intel HPC Dev. Conf.*, Nov. 2019. 153

[KSA+15] R. Komuravelli, M. Sinclair, J. Alsop, et al. Stash: have your scratchpad and cache it too. *ISCA*, Oct. 2015. 137

[KMY+17] J. Konecny, H. McMahan, F. Yu, P. Richtarik, A. Suresh, and D. Bacon. Federated learning: strategies for improving communication efficiency. Oct. 2017. 105

[KCV+20] A. Kosson, V. Chiley, A. Venigalla, J. Hestness, and U. Koster. Pipelined back-propagation at scale: training large models without batches. Mar. 2020. 103, 154

[KWW+17] U. Koster, T. Webb, X. Wang, et al. Flexpoint: an adaptive numerical format for efficient training of deep neural networks. *NeurIPS*, Dec. 2017. 112, 116

[KL19] W. Kouw and M. Loog. An introduction to domain adaptation and transfer learning. Jan. 2019. 95

[KBC+18] T. Kraska, A. Beutel, E. H. Chi, J. Dean, and N. Polyzotis. The case for learned index structures. Apr. 2018. 199

[KSH12] A. Krizhevsky, I. Sutskever, and G. Hinton. ImageNet classification with deep convolutional neural networks. *NeurIPS*, Dec. 2012. 5, 28, 38, 48, 102

[Kri14] A. Krizhevsky. One weird trick for parallelizing convolutional neural networks. Apr. 2014.

[KGG+18] O. Kuchaiev, B. Ginsburg, I. Gitman, et al. Mixed-precision training for NLP and speech recognition with OpenSeq2Seq. Nov. 2018. 115

[LMM+19] I. Laguna, R. Marshall, K. Mohror, M. Ruefenacht, A. Skjellum, and N. Sultana. A large-scale study of MPI usage in open-source HPC applications. *SC*, Nov. 2019. 106

[LCG+19] Z. Lan, M. Chen, S. Goodman, K. Gimpel, P. Sharma, and R. Soricut. ALBERT: a lite BERT for self-supervised learning of language representations. Sep. 2019. 65

[LS19] R. Larsen and T. Shpeisman. TensorFlow graph optimizations. 2019. 196

[LA04] C. Lattner and V. Adve. LLVM: a compilation framework for lifelong program analysis & transformation. *CGO*, Mar. 2004. 164

[LP19] C. Lattner and J. Pienaar. MLIR primer: a compiler infrastructure for the end of Moore's Law. *CGO*, Feb. 2019. 166

[LG16] A. Lavin and S. Gray. Fast algorithms for convolutional neural networks. *CVPR*, Sep. 2015. 33

[Lec16] Y. Lecun. RI seminar: Yann LeCun : the next frontier in AI: unsupervised learning. Nov. 2016. 14

[LBB+98] Y. LeCun, L. Bottou, Y. Bengio, and P. Haffner. Gradient-based learning applied to document recognition. *IEEE*, Nov. 1998. 17, 50

[LDS89] Y. Lecun, J. Denker, and S. Solla. Optimal brain damage. *NeurIPS*, 1989. 121

[LAG+19] M. Lecuyer, V. Atlidakis, R. Geambasu, D. Hsu, and S. Jana. Certified robustness to adversarial examples with differential privacy. *S&P*, May 2019. 203

[LTH+16] C. Ledig, L. Theis, F. Huszar, et al. Photo-realistic single image super-resolution using a generative adversarial network. *CVPR*, Sep. 2016. 59

[LMC+17] E. Lee, D. Miyashita, E. Chai, B. Murmann, and S. Wong. LogNet: energy-efficient neural networks using logarithmic computation. *ICASSP*, Mar. 2017. 118

[LLH+19] J. Lee, J. Lee, D. Han, J. Lee, G. Park, and H. Yoo. 7.7 LNPU: a 25.3TFLOPS/W sparse deep-neural-network learning processor with fine-grained mixed precision of FP8-FP16. *ISSCC*, Feb. 2019. 117, 122, 150

[LMR+19] K. Lee, S. Maji, A. Ravichandran, and S. Soatto. Meta-learning with differentiable convex optimization. *CVPR*, Apr. 2019. 200

[LLX+20] D. Lepikhin, H. Lee, Y. Xu, et al. GShard: scaling giant models with conditional computation and automatic sharding. June 2020. 99, 102, 107

[LAS+07] J. Leverich, H. Arakida, A. Solomatnikov, A. Firoozshahian, M. Horowitz, and C. Kozyrakis. Comparing memory systems for chip multiprocessors. *ISCA*, June 2007. 136

[LM18] Y. Leviathan and Y. Matias. Google Duplex: an AI system for accomplishing real-world tasks over the phone. May 2018. 68

[LSZ+19] T. Li, A. Sahu, M. Zaheer, M. Sanjabi, A. Talwalkar, and V. Smith. Federated optimization in heterogeneous networks. Sep. 2019. 104, 105

[LCH+19] X. Li, S. Chen, X. Hu, and J. Yang. Understanding the disharmony between dropout and batch normalization by variance shift. *CVPR*, Jan. 2019. 41

[LKH+18] D. Liang, R. Krishnan, M. Hoffman, and T. Jebara. Variational autoencoders for collaborative tiltering. *IW3C2*, Feb. 2018. 46

[LHP+19] T. Lillicrap, J. Hunt, A. Pritzel, N. Heess, T. Erez, Y. Tassa, D. Silver, and D. Wierstra. Continuous control with deep reinforcement learning. July 2019. 71

[LGH19] J. Lin, C. Gan, and S. Han. Defensive quantization: when efficiency meets robustness. *ICLR*, Apr. 2019. 123, 202

[LGH+16] T. Lin, P.Dollár, R. Girshick, K. He, B. Hariharan, and S. Belongie. Feature pyramid networks for object detection. *CVPR*, Dec. 2016. 55

[LGG+17] T. Lin, P. Goyal, R. Girshick, K. He, and P. Dollar. Focal loss for dense object detection. *ICCV*, Aug. 2017. 56

[LSP+19] T. Lin, S. Stich, K. Patel, and M. Jaggi. Don't use large mini-batches, use local SGD. June 2019.

[LHM+18] Y. Lin, S. Han, H. Mao, Y. Wang, and W. Dally. Deep gradient compression: reducing the communication bandwidth for distributed training. *ICLR*, Feb. 2018. 101

[LHL+18] P. Lindstrom, J. Hittinger, M. Larsen, S. Lloyd, and M. Salasoo. Alternatives to IEEE: NextGen number formats for scientific computing. *IPAM*, Oct. 2018. 128

[LRS+18] G. Liu, F. Reda, K. Shih, T. Wang, A. Tao, and B. Catanzaro. Image inpainting for irregular holes using partial convolutions. *ECCV*, Apr. 2018. 30

[LDR+18] L. Liu, S. Dean, E. Rolf, M. Simchowitz, and M. Hardt. Delayed impact of fair machine learning. *ICML*, Apr. 2018. 205

[LPH+18] X. Liu, J. Pool, S. Han, and W. Dally. Efficient sparse Winograd convolutional neural networks. *ICLR*, Feb. 2018. 33

[LSY19] H. Liu, K. Simonyan, and Y. Yang. DARTS: differentiable architecture search. *ICLR*, Apr. 2019. 200

[LJH+19] L. Liu, H. Jiang, P. He, W. Chen, X. Liu, J. Gao, and J. Han. On the variance of the adaptive learning rate and beyond. Aug. 2019. 85

[LZL+19] L. Liu, J. Zhu, Z. Li, Y. Lu, Y. Deng, J. Han, S. Yin, and S. Wei. A survey of coarse-grained reconfigurable architecture and design: taxonomy, challenges, and applications. *CSUR*, Oct. 2019. 146

[LAE+15] W. Liu, D. Anguelov, D. Erhan, C. Szegedy, S. Reed, C. Fu, and A. Berg. SSD: single shot multibox detector. *ECCV*, Dec. 2015. 56

[LOG+19] Y. Liu, M. Ott, N. Goyal, et al. RoBERTa: a robustly optimized BERT pretraining approach. July 2019.

[LSZ+19] Z. Liu, M. Sun, T. Zhou, G. Huang, and T. Darrell. Rethinking the value of network pruning. *ICLR*, Mar. 2019. 122

[Llv20] LLVM. MLIR: the case for a simplified polyhedral form. 2020. 196

[LSD14] J. Long, E. Shelhamer, and T. Darrell. Fully convolutional networks for semantic segmentation. *CVPR*, Nov. 2014. 57

[Lor19] B. Lorica. One simple graphic: researchers love PyTorch and TensorFlow. July 2019. 182, 189

[LH17] I. Loshchilov and F. Hutter. SGDR: stochastic gradient descent with warm restarts. *ICLR*, May 2017. 83

[LH19] I. Loshchilov and F. Hutter. Decoupled weight decay regularization. *ICLR*, Jan. 2019. 84, 85

[Lov19] S. Lovely. How many titles are available on Netflix in your country?. May 2019. 44

[LPM15] M. Luong, H. Pham, and C. Manning. Effective approaches to attention-based neural machine translation. Aug. 2015. 63

[LCZ+19] S. Lym, E. Choukse, S. Zangeneh, W. Wen, S. Sanghavi, and M. Erez. PruneTrain: fast neural network training by dynamic sparse model reconfiguration. *SC*, Nov. 2019. 123

[MYM+19] L. Ma, Z. Yang, Y. Miao, J. Xue, M. Wu, L. Zhou, and Y. Dai. NeuGraph: parallel deep neural network computation on large graphs. *ATC*, July 2019. 13, 47

[MMS+19] A. Madry, A. Makelov, L. Schmidt, D. Tsipras, and A. Vladu. Towards deep learning models resistant to adversarial attacks. *ICLR*, Sep. 2019. 202

[MHP+17] H. Mao, S. Han, J. Pool, W. Li, X. Liu, Y. Wang, and W. Dally. Exploring the regularity of sparse structure in convolutional neural networks. *NeurIPS*, Dec. 2017. 122

[ML18] D. Masters and C. Luschi. Revisiting small batch training for deep neural networks. Apr. 2018. 76, 77, 80, 93

[MKA+18] S. McCandlish, J. Kaplan, D. Amodei, et al. An empirical model of large-batch training. Feb. 2017. 93

[MMR+17] H. McMahan, E. Moore, D. Ramage, S. Hampson, and B. Arcas. Communication-efficient learning of deep networks from decentralized data. Feb. 2017. 103

[MSD+19] N. Mellempudi, S. Srinivasan, D. Das, and B. Kaul. Mixed precision training with 8-bit floating point. May 2019. 115, 117

[MC17] D. Meng and H. Chen. MagNet: a two-pronged defense against adversarial examples. *CCS*, Sep. 2017. 202

[Mer19] S. Merity. Single headed attention RNN: stop thinking with your head. Nov. 2019. 62

[Met19] MetaFlow. A framework for real-life data science. 2019. 201

[Met19b] Metaflow. Metaflow on AWS. 2019. 201

[MLN19] P. Michel, O. Levy, and G. Neubig. Are sixteen heads really better than one?. *NeurIPS*, Dec. 2019. 65

[Mic20] Microsoft. ML.NET Documentation. 2020. 201

[Mic20b] Microsoft. Azure Cognitive services. 2020. 201

[Mig17] S. Migacz. 8-bit inference with TensorRT. *GTC*, May 2017. 120

[MSU+19] H. Mikami, H. Suganuma, P. U-chupala, Y. Tanaka, and Y. Kageyama. Massively distributed SGD: ImageNet/ResNet-50 training in a flash. Mar. 2019. 102

[MSC+13] T. Mikolov, I. Sutskever, K. Chen, G. Corrado, and J. Dean. Distributed representations of words and phrases. and their compositionality. *NeurIPS*, Dec. 2013. 38

[MNA16] F. Milletari, N. Navab, and S. Ahmadi. V-Net: fully convolutional neural networks for volumetric medical image segmentation. *3DV*, June 2016. 57

[MGP+18] A. Mirhoseini, A. Goldie, H. Pham, B. Steiner, Q. Le, and J. Dean. A hierarchical model for device placement. *ICLR*, 2018. 102, 199

[MFL+19] S. Mirzadeh, M. Farajtabar, A. Li, N. Levine, A. Matsukawa, and H. Ghasemzadeh. Improved knowledge distillation via teacher assistant. *AAAI*, Dec. 2019. 124

[MWZ+19] M. Mitchell, S. Wu, A. Zaldivar, et al. Model cards for model reporting. Jan. 2019. 205

[MZH+16] I. Mitliagkas, C. Zhang, S. Hadjis, and C. Re. Asynchrony begets momentum, with an application to deep learning. *Comm., Control, and Comp.*, Nov. 2016. 101

[Mlf20] MLFlow. An open source platform for the machine learning lifecycle. 2020. 202

[Mlp18] MLPerf. 2018. 157

[MBM+16] V. Mnih, A. Badia, M. Mirza, et al. Asynchronous methods for deep reinforcement learning. *ICML*, June 2016. 71

[KSe+13] V. Mnih, K. Kavukcuoglu, D. Silver, et al. Playing atari with deep reinforcement learning. Dec. 2013. 71

[MKS+15] V. Mnih, K. Kavukcuoglu, D. Silver, et al. Human-level control through deep reinforcement learning. *Nature*, Feb. 2015. 70, 71

[Moo65] G. Moore. Cramming more components onto integrated circuits. *Electronics*, Apr. 1965. 131

[Moo75] G. Moore. Progress in digital integrated electronics. *Technical Digest*, Sep. 1975. 131

[MPG+20] R. Mor, E. Peterfreund, M. Gavish, and A. Globerson. Optimal strategies against generative attacks. *ICLR*, Feb. 2020. 203

[MYP+19] A. Morcos, H. Yu, M. Paganini, and Y. Tian. One ticket to win them all: generalizing lottery ticket initializations across datasets and optimizers. *NeurIPS*, Dec. 2019. 199

[Mor19] T. Morgan. Nvidia shows off tech chops with RC18 inference chip. *Next Platform*, Sep. 2019. 152

[MNW+18] P. Moritz, R. Nishihara, S. Wang, et al. Ray: a distributed framework for emerging AI applications. *OSDI*, Sep. 2018. 70

[Mos17] R. Mosic. Deep reinforcement learning based trading application at JP Morgan Chase. July 2017. 70

[MY17] T. Munkhdalai and H. Yu. Meta networks. *ICML*, June 2017. 200

[NvB+19] M. Nagel, M. van Baalen, T. Blankevoort, and M. Welling. Data-free quantization through weight equalization and bias correction. *CVPR*, Nov. 2019. 121

[NIG+18] D. Nagy, G. Indalecio, A. Garcia-Loureiro, M. Elmessary, K. Kalna, and N. Seoane. FinFET versus gate-all-around nanowire FET: performance, scaling, and variability. *EDS*, Feb. 2018.

[Nak19] P. Nakkiran. Adversarial robustness may be at odds with simplicity. adversarial robustness may be at odds with simplicity. Jan. 2019. 202

[NKB+20] P. Nakkiran, G. Kaplun, Y. Bansal, T. Yang, B. Barak, and I. Sutskever. Deep double descent: where bigger models and more data hurt. *ICLR*, Apr. 2020. 75, 76

[Nar19] N. Narayanan. How to recognize AI snake oil. 2019. 2, 206

[NSA+19] A. Nassif, I. Shahin, I. Attili, M. Azzeh, and K. Shaalan. Speech recognition using deep neural networks: a systematic review. *Access*, 2019. 65

[NMS+19] M. Naumov, D. Mudigere, H. Shi, et al. Deep learning recommendation model for personalization and recommendation systems. May 2019. 47, 48

[NKM+20] M. Naumov, J. Kim, D. Mudigere, et al. Deep learning training in Facebook data centers: design of scale-up and scale-out systems. Mar. 2020. 3, 102, 152, 156

[Nay19] P. Nayak. Understanding searches better than ever before. Oct. 2019. 64

[NMZ19] E. Neftci, H. Mostafa, and F. Zenke. Surrogate gradient learning in spiking neural networks: bringing the power of gradient-based optimization to spiking neural networks. *SPM*, Nov. 2019. 15

[Nea95] R. Neal. Bayesian learning for neural networks. Ph.D. Thesis, University of Toronto, 1995. 15, 204

[Nim20] Nimbix. Groq tensor streaming processors. 2020. 154

[NDC+17] J. Novikova, O. Dusek, A. Curry, and V. Rieser. Why we need new evaluation metrics for NLG. July 2017. 61

232 BIBLIOGRAPHY

[NKJ+19] E. Nurvitadhi, D. Kwon, A. Jafari, et al. Why compete when you can work together: FPGA-ASIC integration for persistent RNNs. *FCCM*, May 2019. 153

[Nvi15] Nvidia. PTX and SASS assembly debugging. 2015. 144

[Nvi20] Nvidia. RAPIDS. 2020. 202

[Nvi20b] Nvidia. T4. 2020. 134

[Nvi20c] Nvidia. Data center deep learning product performance. July 2020. 99

[OSJ+18] C. Olah, A. Satyanarayan, I. Johnson, S. Carter, L. Schubert, K. Ye, and A. Mordvintsev. The building blocks of interpretability. 2018. 204

[OPM02] T. Ojala, M. Pietikäinen, and T. Maenpaa. Multiresolution gray-scale and rotation invariant texture classification with local binary patterns. *PAMI*, July 2002. 48

[Ope18] OpenAI. Kinds of RL algorithms. 2018. 71

[Orr99] G. Orr. Momentum and Learning Rate Adaptation. Willamette University, 1999. 83

[Pad19] S. Padmanabhan. Building a product catalog: eBay's university machine learning competition. Oct. 2019. 44

[PdO+18] M. Paganini, L. de Oliveira, and B. Nachman. Accelerating science with generative adversarial. networks: an application to 3D particle showers in multi-layer calorimeters. *PRL*, Jan. 2018. 14

[PY10] S. Pan and Q. Yang. A survey on transfer learning. *TKDE*, Oct. 2010. 95

[PMW+16] N. Papernot, P. McDaniel, X. Wu, S. Jha, and A. Swami. Distillation as a defense to adversarial perturbations against deep neural networks. *S&P*, Mar. 2016. 202

[PCZ+19] D. Park, W. Chan, Y. Zhang, C. Chiu, B. Zoph, E. Cubuk, and Q. Le. SpecAugment: a simple data augmentation method for automatic speech recognition. Apr. 2019. 67, 89

[PNB+18] J. Park, M. Naumov, P. Basu, S. Deng, et al. Deep learning inference in Facebook data centers: characterization, performance optimizations and hardware implications. Nov. 2018. 54, 117

[PRH+17] A. Pedram, S. Richardson, M. Horowitz, S. Galal, and S. Kvatinsky. Dark memory and accelerator-rich system optimization in the dark silicon era. *D&T*, May 2016. 138

[PSC+19] M. Pellauer, Y. Shao, J. Clemons, et al. Buffets: an efficient and composable storage idiom for explicit decoupled data orchestration. *ASPLOS*, Apr. 2019. 137

[PSM14] J. Pennington, R. Socher, and C. Manning. GloVe: global vectors for word representation. *EMNLP*, 2014. 38

[PGZ+18] H. Pham, M. Guan, B. Zoph, Q. V. Le, and J. Dean. Efficient neural architecture search via parameter sharing. Feb. 2018. 200

[Phi18] M. Phi. Illustrated guide to LSTM's and GRU's: a step by step explanation. *TDS*. Sep. 2018. 36

[PPG+17] W. Ping, K. Peng, A. Gibiansky, S. Arik, A. Kannan, S. Narang, J. Raiman, and J.Miller. Deep Voice 3: scaling text-to-speech with convolutional sequence learning. Oct. 2017. 68

[PPC18] W. Ping, K. Peng, and J. Chen. ClariNet: parallel wave generation in end-to-end text-to-speech. July 2018. 69

[Pol99] F. Pollack. New microarchitecture challenges in the coming generations of CMOS process technologies. *MICRO*, Nov. 1999.

[PZK+17] R. Prabhakar, Y. Zhang, D. Koeplinger, et al. Plasticine: a reconfigurable architecture for parallel patterns. *SIGARCH*, June 2017. 154

[PHX+18] V. Pratap, A. Hannun, Q. Xu, et al. wav2letter++: the fastest open-source speech recognition system. Dec. 2018. 67

[Qia99] N. Qian. On the momentum term in gradient descent learning algorithms. Jan. 1999. 83

[RMC15] A. Radford, L. Metz, and S. Chintala. Unsupervised representation learning with deep convolutional generative adversarial networks. *ICIGP*, Nov. 2015. 59

[RWC+19] A. Radford, J. Wu, R. Child, D. Luan, D. Amodei, and I. Sutskever. Language models are unsupervised multitask learners. 2019.

[RBA+13] J. Ragan-Kelley, C. Barnes, A. Adams, S. Paris, F. Durand, and S. Amarasinghe. Halide: a language and compiler for optimizing parallelism, locality, and recomputation in image processing pipelines. *PLDI*, June 2013. 196

[RSR+19] C. Raffel, N. Shazeer, A. Roberts, et al. Exploring the limits of transfer learning with a unified text-to-text transformer. Oct. 2019. 102

[RZQ+19] K. Rakelly, A. Zhou, D. Quillen, C. Finn, and S. Levine. Efficient off-policy meta-reinforcement learning via probabilistic context variables. Mar. 2019. 199

[ROR+16] M. Rastegari, V. Ordonez, J. Redmon, and A. Farhadi. XNOR-Net: ImageNet classification using binary convolutional neural networks. *ECCV*, Sep. 2016. 119

[RD19] S. Raza and C. Ding. Progress in context-aware recommender systems–an overview. Jan. 2019. 44

[RAH+19] E. Real, A. Aggarwal, Y. Huang, and Q. Le. Regularized evolution for image classifier architecture search. *AAAI*, Feb. 2019. 200

[RKK19] S. Reddi, S. Kale, and S. Kumar. On the convergence of Adam and beyond. *ICLR*, Apr. 2019. 84, 85

[RDG+16] J. Redmon, S. Divvala, R. Girshick, and A. Farhadi. You only look once: unified, real-time object detection. *CVPR*, 2016. 55

[RF18] J. Redmon and A. Farhadi. YOLOv 3: an incremental improvement. Apr. 2018. 56

[RHG+15] S. Ren, K. He, R. Girshick, and J. Sun. Faster R-CNN: towards real-time object detection with region proposal networks. *NeurIPS*, Dec. 2015. 55

[RAA+19] C. Renggli, S. Ashkboos, M. Aghagolzadeh, D. Alistarh, and T. Hoefler. SparCML: high-performance sparse communication for machine learning. *SC*, Aug. 2019. 107

[RKL+18] A. Rodriguez, T. Kacprzak, A. Lucchi, et al. Fast cosmic web simulations with generative adversarial networks. *CompAC*, Nov. 2018. 14

[RKB+09] B. Rogers, A. Krishna, G. Bell, K. Vu, X. Jiang, and Y. Solihin. Scaling the bandwidth wall: challenges in and avenues for CMP scaling. *SIGARCH*, Jun. 2009. 127

[RDK+19] D. Rolnick, P. Donti, L. Kaack, et al. Tackling climate change with machine learning. Nov. 2019.

[RDK+19] D. Rolnick, P. Donti, L. Kack, et al. Tackling climate change with machine learning workshop. *NeurIPS*, Dec. 2019. 206

[RFB15] O. Ronneberger, P. Fischer, and T. Brox. U-Net convolutional networks for biomedical image segmentation. May 2015. 57

[Ros20] C. Rosset. Turing-NLG: a 17-billion-parameter language model by Microsoft. Feb. 2020.

[RXT19] B. Roune and XLA Team. Compiling ML with XLA. Feb. 2019.

[RJP19] K. Roy, A. Jaiswal, and P. Panda. Towards spike-based machine intelligence with neuromorphic computing. *Nature*, 2019. 15, 150

[Rud17] S. Ruder. An overview of multi-task learning in deep neural networks. June 2017. 95

[Rup20] K. Rupp. Microprocessor trend data. 2020. 135

[RDS+15] O. Russakovsky, J. Deng, H. Su, et al. Large scale visual recognition challenge. *IJCV*, 2015. 48

[RRS+19] A. Rusu, D. Rao, J. Sygnowski, O. Vinyals, R. Pascanu, S. Osindero, and R. Hadsell. Meta-learning with latent embedding optimization. *ICLR*, Mar. 2019. 200

[Sam16] Samgsung. Samsung begins mass producing world's fastest DRAM–based on newest high bandwidth memory (HBM) interface. 2016. 138

[SST09] P. Sanders, J. Speck, and J. Traff. Two-tree algorithms for full bandwidth broadcast, reduction and scan. Sep. 2009. 109

[SDC+19] V. Sanh, L. Debut, J. Chaumond, and T. Wolf. DistilBERT, a distilled version of BERT: smaller, faster, cheaper and lighter. Oct. 2019. 65

[San19] V. Sanh. Smaller, faster, cheaper, lighter: introducing DistilBERT, a distilled version of BERT. *Medium*, Aug. 2019. 66

[Sas19] K. Sasaki. Federated Learning with TensorFlow. 2019. 104

[SYP17] K. Sato, C. Young, and D. Patterson. An in-depth look at Google's first Tensor Processing Unit (TPU). May 2017. 174

[SGT+09] F. Scarselli, M. Gori, A. Tsoi, M. Hagenbuchner, and G. Monfardini. The graph neural network model. *TNNLS*, Jan. 2009. 13

[Sch19] J. Schalkwyk. An all-neural on-device speech recognizer. Mar. 2019. 67

[SAH+20] J. Schrittwieser, I. Antonoglou, T. Hubert, et al. Mastering Atari, Go, Chess and Shogi by planning with a learned model. Feb. 2020. 72, 199

[SKP15] F. Schroff, D. Kalenichenko, and J. Philbin. FaceNet: a unified embedding for face recognition and clustering. *CVPR*, Mar. 2015. 59

[SLM+17] J. Schulman, S. Levine, P. Moritz, M. Jordan, and P. Abbeel. Trust region policy optimization. Apr. 2017. 71

[SFD+14] F. Seide, H. Fu, J. Droppo, G. Li, and D. Yu. 1-bit stochastic gradient descent and application to data-parallel distributed training of speech DNNs. *Int' Speech Comm. Association*, Sep. 2014. 101

[SDB18] A. Sergeev and M. Del Balso. Horovod: fast and easy distributed deep learning in TensorFlow. Feb. 2018. 107

[SHB15] R. Sennrich, B. Haddow, and A. Birch. Neural machine translation of rare words with subword units. Aug. 2015. 61, 62

[SKF+16] M. Seo, A. Kembhavi, A. Farhadi, and H. Hajishirzi. Bidirectional attention flow for machine comprehension. Nov. 2016. 62

[SLA+19] C. Shallue, J. Lee, J. Antognini, J. Sohl-Dickstein, R. Frostig, and G. Dahl. Measuring the effects of data parallelism on neural network training. *JMLR*, July 2019. 82, 83

[SWR18] Y. Sharan, H. Wang, and S. Rath. GUI testing powered by deep learning. *eBay Tech Blog*. June 2018.

[SCP+18] N. Shazeer, Y. Cheng, N. Parmar, et al. Mesh-TensorFlow: deep learning for supercomputers. *NeurIPS*, Dec. 2018. 102

[SPW+17] J. Shen, R. Pang, R. Weiss, et al. Natural TTS synthesis by conditioning WaveNet on Mel Spectrogram predictions. *ICASSP*, Dec. 2017. 68

[SDY+19] S. Shen, Z. Dong, J. Ye, L. Ma, Z. Yao, A. Gholami, M. Mahoney, and K. Keutzer. Q-BERT: Hessian based ultra low precision quantization of BERT. Sep. 2019. 65, 117

[She18] R. Sheth. Introducing PyTorch across Google Cloud. Oct. 2018. 190

[SLA+19] B. Shickel, T. Loftus, L. Adhikari, T. Ozrazgat-Baslanti, A. Bihorac, and P. Rashidi. DeepSOFA: a continuous acuity score for critically ill patients using clinically interpretable deep learning. Feb. 2019. 204

[SPP+19] M. Shoeybi, M. Patwary, R. Puri, P. LeGresley, J. Casper, and B. Catanzaro. Megatron LM training multi billion parameter language models using model parallelism. Oct. 2019. 3, 99, 102

[SL19] T. Shpeisman and C. Lattner. MLIR: multi-level intermediate representation for compiler infrastructure. Apr. 2019. 184, 195

[SHM+16] D. Silver, A. Huang, C. Maddison, et al. Mastering the game of Go with deep neural networks and tree search. *Nature*, Jan. 2016. 44, 72

[SSS+17] D. Silver, J. Schrittwieser, K. Simonyan, I. Antonoglou, et al. Mastering the game of Go without human knowledge. *Nature*, Oct. 2017. 72

[SSS+18] D. Silver, J. Schrittwieser, K. Simonyan, et al. A general reinforcement learning algorithm that masters chess, shogi, and Go through self-play. *Science*, Dec. 2018. 72

[SZ14] K. Simonyan and A. Zisserman. Very deep convolutional networks for large-scale image recognition. Sep. 2014. 50

[Smi17] L. Smith. Cyclical learning rates for training neural networks. *WACV*, Apr. 2017. 83, 93

[SSZ17] J. Snell, K. Swersky, and R. Zemel. Prototypical networks for few-shot learning. *NeurIPS*, Dec. 2017. 200

[Ste19] I. Steinwart. A sober look at neural network initializations. Sep. 2019. 79

[Ste19b] N. Stephens. BFloat16 processing for neural networks on Armv8-A. Aug. 2019. 152

[SA19] A. Stooke and P. Abbeel. Accelerated methods for deep reinforcement learning. Jan. 2019. 70

[SPE19] A. Straw, A. Procter, and R. Earhart. nGraph: unlocking next-generation performance with deep learning compilers. 2019. 168, 169, 196, 197

[SGB+19] S. Sukhbaatar, E. Grave, P. Bojanowski, and A. Joulin. Adaptive attention span in transformers. May 2019. 65

[SCC+19] X. Sun, J. Choi, C. Chen, et al. Hybrid 8-bit floating point (HFP8) training and inference for deep neural networks. *NeurIPS*, Dec. 2019. 117, 119

[SWL+19] Y. Sun, S. Wang, Y. Li, S. Feng, H. Tian, H. Wu, and H. Wang. ERNIE 2.0: a continual pre-training framework for language understanding. 2019.

[SAD+20] Y. Sun, N. Agostini, S. Dong, and D. Kaeli. Summarizing CPU and GPU design trends with product data. 2020. 6

[SVL14] I. Sutskever, O. Vinyals, and Q. Le. Sequence to sequence learning with neural networks. *NeurIPS*, Dec. 2014. 62

[SCY+17] V. Sze, Y. Chen, T. Yang, and J. Emer. Efficient processing of deep neural networks: a tutorial and survey. *Proc. IEEE*, Dec. 2017. 148, 149

[SCY+20] V. Sze, Y. Chen, T. Yang, and J. Emer. Efficient processing of deep neural networks. *M&C*, June 2020. 147

[SLJ+14] C. Szegedy, W. Liu, Y. Jia, et al. Going deeper with convolutions. *CVPR*, Sep. 2014. 51

[SVI+15] C. Szegedy, V. Vanhoucke, S. Ioffe, J. Shlens, and Z. Wojna. Rethinking the Inception architecture for computer vision. *CVPR*, Dec. 2015. 51, 77

[SZS+14] C. Szegedy, W. Zaremba, I. Sutskever, J. Bruna, D. Erhan, I. Goodfellow, and R. Fergus. Intriguing properties of neural networks. Feb. 2014. 202

[Syn17] Synced. A brief overview of attention mechanism. *Medium*, Sep. 2017. 40

[TPL19] M. Tan, R. Pang, and Q. Le. EfficientDet: scalable and efficient object detection. Nov. 2019. 56, 200

[TL19]   M. Tan and Q. Le. EfficientNet: rethinking model scaling for convolutional neural networks. May 2019. 54, 200

[TYD+18]   Y. Tassa, Y, Doron, A. Muldal, et al. DeepMind control suite. Jan. 2018. 70

[TKT+16]   S. Tavarageri, W. Kim, J. Torrellas, and P. Sadayappan. Compiler support for software cache coherence. *HiPC*, Dec. 2016. 134

[Ter19]   Terry. Inlining decisions in visual studio. July 2019.

[TRG05]   R. Thakur, R. Rabenseifner, and W. Gropp. Optimization of collective communication operations in MPICH. *HiPC*, Feb. 2005. 106

[TGL+20]   N. Thompson, K. Greenewald, K. Lee, and G. Manso. The computational limits of deep learning. July 2020. 1

[TKP+18]   F. Tramer, A. Kurakin, N. Papernot, I. Goodfellow, D. Boneh, and P. McDaniel. Ensemble adversarial training: attacks and defenses. *ICLR*, July 2018. 202

[TAN+18]   H. Tsai, S. Ambrogio, P. Narayanan, R. Shelby, and G. Burr. Recent progress in analog memory-based accelerators for deep learning. *J. Phys. D: Appl. Phys*, June 2018. 150

[Tsa18]   S. Tsang. Review: YOLOv1— you only look once (object detection). *TDS*, Oct. 2018. 57

[TSE+19]   D. Tsipras, S. Santurkar, L. Engstrom, A. Turner, and A. Madry. Robustness may be at odds with accuracy. *ICLR*, Sep. 2019. 202

[Tvm19]   TVM. TVM deep learning compiler joins Apache Software Foundation. Mar. 2019. 192

[Tvm19]   TVM. Introduction to Relay IR. 2019. 192

[vKK+16]   A. van den Oord, N. Kalchbrenner, and K. Kavukcuoglu. Pixel recurrent neural networks. Jan. 2016. 59

[vDZ+16]   A. van den Oord, S. Dieleman, H. Zen, et al. WaveNet: a generative model for raw audio. Sep. 2016. 68

[vLB+17]   A. van den Oord, Y. Li, I. Babuschkin, et al. Parallel WaveNet: fast high-fidelity speech synthesis. Nov. 2017. 68

[VS19]   J. Valin and J. Skoglund. LPCNet: improving neural speech synthesis through linear prediction. *ICASSP*, May 2019. 68

[VZT+18] N. Vasilache, O. Zinenko, T. Theodoridis, et al. Tensor Comprehensions: framework-agnostic high-performance machine learning abstractions. *ICASSP*, May 2019. 196

[VSP+17] A. Vaswani, N. Shazeer, N. Parmar, et al. Attention is all you need. *NeurIPS*, Dec. 2017. 12, 39, 63, 64

[VSZ+19] R. Venkatesan, Y. Shao, B. Zimmer, et al. A 0.11 PJ/OP, 0.32-128 TOPS, scalable multi-chip-module-based deep neural network accelerator designed with a high-productivity VLSI methodology. *HCS*, Aug. 2019. 152

[Vil18] M. Villmow. Optimizing NMT with TensorRT. Mar. 2018. 62

[VTB+14] O. Vinyals, A. Toshev, S. Bengio, and D. Erhan. Show and tell: a neural image caption generator. *CVPR*, Nov. 2014. 63

[VBL+17] O. Vinyals, C. Blundell, T. Lillicrap, K. Kavukcuoglu, and D. Wierstra. Matching networks for one shot learning. *NeurIPS*, Dec. 2017. 200

[VBC+19] O. Vinyals, I. Babuschkin, J. Chung, et al. AlphaStar: mastering the real-time strategy game StarCraft II. Dec. 2019. 70

[VAK19] A. Vladimirov, R. Asai, and V. Karpusenko. Parallel programming and optimization with Intel Xeon Phi coprocessors. Jan. 2019. 172

[Wal13] C. Walsh. Peter Huttenlocher (1931-2013). *Nature*, Oct. 2013. 121

[SMH+18] A.Wang, A. Singh, J. Michael, F. Hill, O. Levy, and S. Bowman. GLUE: a multi-task benchmark and analysis platform for natural language understanding. Apr. 2018. 60

[WYL+20] H. Wang, J. Yang, H. Lee, and S. Han. Learning to design circuits. Jan. 2020. 199

[WCB+18] N. Wang, J. Choi, D. Brand, C. Chen, and K. Gopalakrishnan. Training deep neural networks with 8-bit floating point numbers. *NeurIPS*, Dec. 2018. 121

[WVP+19] G. Wang, S. Venkataraman, A. Phanishayee, J. Thelin, N. Devanur, and I. Stoica. Blink: fast and generic collectives for distributed ML. Oct. 2019. 13, 47, 107, 110

[WYZ+17] J. Wang, L. Yu, W. Zhang, Y. Gong, Y. Xu, B. Wang, P. Zhang, and D. Zhang. IRGAN: a minimax game for unifying generative and discriminative information retrieval models. *SIGIR*, May 2017. 46

[WML+19] Y. Wang, A. Mohamed, D. Le, et al. Transformer-based acoustic modeling for hybrid speech recognition. Oct. 2019. 67

[WYK+19] Y. Wang, Q. Yao, J. Kwok, and L. Ni. Generalizing from a few examples: a survey on few-shot learning. *Comp. Surveys*, May 2019. 95

[WWB19] Y. Wang, G. Wei, and D. Brooks. Benchmarking TPU, GPU, and CPU platforms for deep learning. Oct. 2019. 157

[WSS+17] Y. Wang, R. Skerry-Ryan, D. Stanton, et al. Tacotron: towards end-to-end speech synthesis. Mar. 2017. 68

[WWS+19] Y. Wang, Q. Wang, S. Shi, X. He, Z. Tang, K. Zhao, and X. Chu. Benchmarking the performance and power of AI accelerators for AI training. Nov. 2019. 157, 199

[WSA18] R. Wei, L. Schwartz, and V. Adve. DLVM: a modern compiler infrastructure for deep learning systems. *ICLR*, Apr. 2018. 196

[WWW+16] W. Wen, C. Wu, Y. Wang, Y. Chen, and H. Li. Learning structured sparsity in deep neural networks. *NeurIPS*, Dec. 2016. 122

[WXY+17] W. Wen, C. Xu, F. Yan, C. Wu, Y. Wang, Y. Chen, and H. Li. TernGrad: ternary gradients to reduce communication in distributed deep learning. *NeurIPS*, Dec. 2017. 101

[Wen17] L. Weng. From GAN to WGAN. Aug. 2017. 14

[Wik11] Wikimedia. Kernel Machine.svg. 2011. 5

[Wik12] Wikimedia. Cart-pendulum.svg. 2012. 69

[Wik15] Wikimedia. Typical cnn.png. 2015. 11

[Wik17] Wikimedia. MnistExamples.png. 2017. 18

[Wik18] Wikimedia. Spectrogram-19thC.png. 2018. 67

[Wik19] Wikipedia. Apple A13. 2019. 152

[Wik20] Wikipedia. Authors Guild, Inc. v. Google, Inc. Feb. 2020. 206

[Wik20b] Wikipedia. RankBrain. Feb. 2020.

[WWP09] S. Williams, A. Waterman, and D. Patterson. Roofline: an insightful visual performance model for multicore architectures. *ACM*, Apr. 2009. 140

[WRS+18] A. Wilson, R. Roelofs, M. Stern, N. Srebro, and B. Recht. The marginal value of adaptive gradient methods in machine learning. *NeurIPS*, Dec. 2018. 84

[WZL+19] R. Wilson, C. Zhang, W. Lam, D. Desfontaines, D. Simmons-Marengo, and B. Gipson. Differentially private SQL with bounded user contribution. Nov. 2019. 203

[Win20] P. Winder. Reinforcement Learning: industrial applications of intelligent agents. *O'Reilly*, Nov. 2020. 70

[Wri19] L. Wright. New deep learning optimizer, Ranger synergistic combination of RAdam + LookAhead for the best of both. Aug. 2019. 86, 93

[WZX+16] J. Wu, C. Zhang, T. Xue, W. Freeman, and J. Tenenbaum. Learning a probabilistic latent space of object shapes via 3D generative-adversarial modeling. *NeurIPS*, Dec. 2016. 59

[WSC+16] Y. Wu, M. Schuster, Z. Chen, et al. Google's neural machine translation system: bridging the gap between human and machine translation. Sep. 2016. 62

[WAB+17] C. Wu, A. Ahmed, A. Beutel, A. Smola, and H. Jing. Recurrent recommender networks. *WSDM*, Feb. 2017. 47

[WWF+17] S. Wu, J. Wieland, O. Farivar, and J. Schiller. Automatic alt-text: computer-generated image descriptions for blind users on a social network service. *CSCW*, Feb. 2017. 67

[WH18] Y. Wu and K. He. Group normalization. *ECCV*, Mar. 2018. 37, 38

[WZZ+19] B. Wu, X. Zhou, S. Zhao, X. Yue, and K. Keutzer. SqueezeSegV.2: improved model structure and unsupervised domain adaptation for road-object segmentation from a LiDAR point cloud. *ICRA*, May 2019. 59

[WFB+19] F. Wu, A. Fan, A. Baevski, Y. Dauphin, and M. Auli. Pay less attention with lightweight and dynamic convolutions. Jan. 2019. 65

[WKM+19] Y. Wu, A. Kirillov, F. Massa, W. Lo, and R. Girshick. Detectron2. 2019. 58

[WKM+19] Y. Wu, A. Kirillov, F. Massa, W. Lo, and R. Girshick. Detectron.2: a PyTorch-based modular object detection library. 2019. 58

[Wu19] H. Wu. Low precision inference on GPU. *GTC*, Mar. 2019. 115, 120

[WDZ+19] B. Wu, X. Dai, P. Zhang, et al. FBNet: hardware-aware efficient ConvNet design via differentiable neural architecture search. *CVPR*, May 2019. 200

[WM95] W. Wulf and S. McKee. Hitting the memory wall: implications of the obvious. *SIGARCH*, Mar. 1995. 127

[XYB+19] S. Xi, Y. Yao, K. Bhardwaj, P. Whatmough, G. Wei, and D. Brooks. SMAUG: end-to-end full-stack simulation infrastructure for deep learning workloads. Dec. 2019. 157

[XZZ20] C. Xiao, P. Zhong, and C. Zheng. Enhancing adversarial defense by k-winners-take-all. *ICLR*, Feb. 2020. 28, 202

[XGD+17] S. Xie, R. Girshick, P. Dollar, Z. Tu, and K. He. Aggregated residual transformations for deep neural networks. *CVPR*, July 2017. 54, 55

[Xil19] Xilinx. Versal: the first adaptive compute acceleration platform (ACAP). 2019. 153

[XAT+18] C. Xing, D. Arpit, C. Tsirigotis, and Y. Bengio. A walk with SGD. May 2018. 79, 81

[XEQ17] W. Xu, D. Evans, and Y. Qi. Feature squeezing: detecting adversarial examples in deep neural networks. Dec. 2017. 202

[XLF+18] X. Xu, C. Liu, Q. Feng, H. Yin, L. Song, and D. Song. Neural network-based graph embedding for cross-platform binary code similarity detection. *CCS*, July 2018. 202

[YKT+18] M. Yamazaki, A. Kasagi, A. Tabuchi, et al. Yet another accelerated SGD: ResNet-50 training on ImageNet in 74.7 seconds. Mar. 2019. 99

[Yam12] R. Yampolskiy. Turing test as a defining feature of AI-Completeness. *SCI*, 2012. 60

[YCS17] T. Yang, Y. Chen, and V. Sze. Designing energy-efficient convolutional neural networks using energy-aware pruning. *CVPR*, Apr. 2017. 123

[YDY+19] Z. Yang, Z. Dai, Y. Yang, J. Carbonell, R. Salakhutdinov, and Q. Le. XLNet: generalized autoregressive pretraining for language understanding. *NeurIPS*, Dec. 2019. 64

[YHG+15] Z. Yang, X. He, J. Gao, L. Deng, and A. Smola. Stacked attention networks for image question answering. *CVPR*, Nov. 2015. 39, 63

[YGL+18] Z. Yao, A. Gholami, Q. Lei, K. Keutzer, and M. Mahoney. Hessian-based Analysis of large batch training and robustness to adversaries. *NeurIPS*, Dec. 2018. 80

[YGS+20] Z. Yao, A. Gholami, S. Shen, K. Keutzer, and M. Mahoney. AdaHessian: an adaptive second order optimizer for machine learning. Jun. 2020. 85

[YSE+20] J. Yin, S. Sethumurugan, Y. Eckert, N. Enright Jerger, et al. Experiences with ML-driven design: a NoC case study. *HPCA*, Feb. 2020. 199

[YKC+18] C. Ying, S. Kumar, D. Chen, T. Wang, and Y. Cheng. Image classification at supercomputer scale. *NeurIPS*, Dec. 2018. 13, 47, 102

[YGG17] Y. You, I. Gitman, and B. Ginsburg. Large batch training of convolutional networks. Sep. 2017. 85

[YLR+20] Y. You, J. Li, S. Reddi, et al. Large batch optimization for deep learning: training BERT in 76 minutes. *ICLR*, Jan. 2020. 85, 102

[YZH+18] Y. You, Z. Zhang, C. Hsieh, J. Demmel, and K. Keutzer. ImageNet training in minutes. Jan. 2018. 3, 99, 102

[YAB+18] Y. Yu, M. Abadi, P. Barham, et al. Dynamic control flow in large-scale machine learning. *EUROSYS*, May 2018. 180, 184

[YTL+19] L. Yuan, F. Tay, G. Li, T. Wang, and J. Feng. Revisit knowledge distillation: a teacher-free framework. Sep. 2019. 124

[ZK15] S. Zagoruyko and N. Komodakis. Learning to compare image patches via convolutional neural networks. *CVPR*, June 2015. 58

[ZXL+18] N. Zeghidour, Q. Xu, V. Liptchinsky, N. Usunier, G. Synnaeve, and R. Collobert. Fully convolutional speech recognition. Dec. 2018. 67

[Zei12] M. Zeiler. ADADELTA: an adaptive learning rate method. Dec. 2012. 85

[ZF13] M. Zeiler and R. Fergus. Visualizing and understanding convolutional networks. *ECCV*, Nov. 2013. 49, 204

[ZF13] M. Zeiler and R. Fergus. Stochastic pooling for regularization of deep convolutional neural networks. *ICLR*, May 2013. 34

[ZES+20] A. Zela, T. Elsken, T. Saikia, Y. Marrakchi, T. Brox, and F. Hutter. Understanding and robustifying differentiable architecture search. *ICLR*, Jan. 2020. 200

[ZB19] T. Zerrell and J. Bruestle. Stripe: tensor compilation via the nested polyhedral model. Mar. 2019. 193

[ZDH19] B. Zhang, A. Davoodi, and Y. Hu. Efficient inference of CNNs via channel pruning. Aug. 2019. 122

[ZYY18] J. Zhang, J. Yang, and H. Yuen. Training with low-precision embedding tables. *NeurIPS*, Dec. 2018. 115

[ZRW+18] M. Zhang, S. Rajbhandari, W. Wang, and Y. He. DeepCPU: serving RNN-based deep learning models 10x faster. *ATC*, 2018. 62, 134

[ZLH+19] M. Zhang, J. Lucas, G. Hinton, and J. Ba. Lookahead optimizer: k steps forward, 1 step back. *NeurIPS*, Dec. 2019. 85, 86, 153

[ZL19] W. Zhang and P. Li. Spike-train level backpropagation for training deep recurrent spiking neural networks. *NeurIPS*, Dec. 2019. 150

[ZZL+17] X. Zhang, X. Zhou, M. Lin, and J. Sun. ShuffleNet: an extremely efficient convolutional neural network for mobile devices. *CVPR*, July 2017. 59

[ZXH+17] Y. Zhang, T. Xiang, T. Hospedales, and H. Lu. Deep mutual learning. *CVPR*, Jan. 2018. 124

[ZZZ+19] C. Zhao, S. Zhao, M. Zhao, Z. Chen, C. Gao, H. Li, and Y. Tan. Secure multiparty computation: theory, practice and applications. *Inf. Sciences*, Feb. 2019. 203

[ZZX+19] W. Zhao, J. Zhang, D. Xie, Y. Qian, R. Jia, and P. Li. AIBox: CTR prediction model training on a single node. *CIKM*, Nov. 2019. 45

[ZHW+19] Z. Zhao, L. Hong, L. Wei, et al. Recommending what video to watch next: a multitask ranking system. *RecSys*, Sep. 2019. 46

[ZZZ+18] G. Zheng, F. Zhang, Z. Zheng, Y. Xiang, N. Yuan, X. Xie, and Z. Li. DRN: a deep reinforcement learning framework for news recommendation. *IW3C2*, Apr. 2018. 46, 153

[ZMF+18] G. Zhou, N. Mou, Y. Fan, Q. Pi, W. Bian, C. Zhou, X. Zhu, and K. Gai. Deep interest evolution network for click-through rate prediction. *AAAI*, Nov. 2018. 47

[ZTZ+18] Z. Zhuang, M. Tan, B. Zhuang, J. Liu, Y. Guo, Q. Wu, J. Huang, and J. Zhu. Discrimination aware channel pruning for deep neural networks. *NeurIPS*, Dec. 2018. 122

[ZZY+19] R. Zhu, K. Zhao, H. Yang, W. Lin, C. Zhou, B. Ai, Y. Li, and J. Zhou. AliGraph: a comprehensive graph neural network platform. *PVLDB*, Aug. 2019. 13, 47

[Zis18] A. Zisserman. Self-supervised learning. July 2018. 49

[ZL17] B. Zoph and Q. Le. Neural architecture search with reinforcement learning. Feb. 2017. 199, 200

Author's Biography

ANDRES RODRIGUEZ

Andres Rodriguez is a Sr. Principal Engineer and AI Architect in the Data Platform Group at Intel Corporation where he designs deep learning solutions for Intel's customers and provides technical leadership across Intel for deep learning hardware and software products. He has 15 years of experience working in AI. Andres received a Ph.D. from Carnegie Mellon University for his research in machine learning. He was the lead instructor in the Coursera course An Introduction to Practical Deep Learning to over 20 thousand students. He has been an invited speaker at several AI events, including AI with the Best, ICML, CVPR, AI Frontiers Conference, Re-Work Deep Learning Summit, TWIML, Startup MLConf, Open Compute Platform Global Summit, AWS re:Invent, Baidu World, Baidu Cloud ABC Inspire Summit, Google Cloud OnAir Webinar, and several Intel events, as well as an invited lecturer at Carnegie Mellon University, Stanford University, UC Berkeley, and Singularity University.

Printed in the United States
by Baker & Taylor Publisher Services